BREAKING LIFE SCIENCES VOL III

THE UNIVERSAL CONTROL LAYER

The Biological Software of Spaceship Earth - and the Off-World Imperative

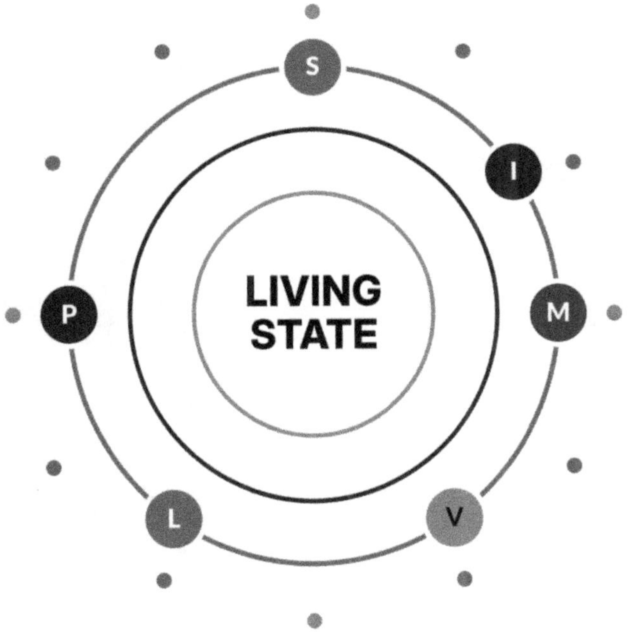

Ross O. Youngs

First-Edition | September 2026

For permission requests or publisher information, contact:
Ross Youngs (Author/Publisher)
2545 Farmers Dr., Suite 370
Columbus, OH 43235
https://rossyoungs.com/
ISBN 979-8-9973605-0-4 (Hardcover)
ISBN 979-8-9973605-1-1 (eBook – Kindle)
Library of Congress Control Number: 2026920057
(Library of Congress Cataloging-in-Publication Data is available)
First Edition
This book is a work of nonfiction. It combines documented historical events, established scientific findings, the author's synthesis of those findings, an engineering architecture proposed for further development, and future scenarios that remain speculative. It is not medical advice and should not be used to diagnose, prevent, or treat any disease.

References to Biosortia and Industrial-Scale Microbiome Mining reflect the author's perspective and firsthand experience; they do not substitute for independent scientific validation. This work intentionally omits confidential source coordinates, undisclosed molecular structures, proprietary algorithms, unpublished assay details, and other protectable information.

Historical scenes are reconstructed from public records. Unless quotation marks and a source are provided, dialogue is paraphrased or omitted. Personal scenes are drawn from the author's previously documented recollections.

Printed in the United States of America

Dedication

To the people in my life who have supported my efforts, taught me, and put up with my occasional lack of work life balance.

To those responsible for keeping a living system alive -

and to those who understand that the system is always answering back.

Contents

Narrative Architecture

Ten Systems, One Control Problem

Each chapter begins with a system that appeared to be under control - until biology exposed what the operators had not measured.

1. **Biosphere 2 - The Oxygen That Disappeared.** A sealed world revealed that an inventory of parts is not a model of living state.
2. **Fordlandia - The Plantation That Invited Its Enemy.** Industrial order removed the ecological spacing that had been doing invisible protective work.
3. **Legionnaires' Disease - The Hotel That Became a Habitat.** A building designed for people became an amplification system for a microbe no one knew to look for.
4. **PKU Newborn Screening - The Baby Who Looked Healthy.** A few drops of blood demonstrated that the best moment to intervene can arrive before symptoms.
5. **The NASA Twins Study - One Twin Left Earth.** A human baseline became a moving trajectory rather than a single normal range.
6. **Mir and the International Space Station - The Biology Behind the Panels.** A spacecraft became an ecosystem whose inhabitants included crew, microbes, condensation, polymers, and time.
7. **The Panama Canal - The Canal Medicine Built.** Mechanical scale became possible only after the disease system was changed upstream.
8. **HIV Combination Therapy - The Cocktail That Bent the Curve.** A rapidly adapting network yielded when medicine stopped relying on one point of attack.
9. **Recombinant Insulin and Portable Biomanufacturing - The Message That Became Medicine.** Biological information became a manufacturing instruction, opening a path from data to matter.

10. **Apollo 7 - The Cold That Rewrote the Flight Plan.** A technically successful mission revealed that the human being is not outside the control system.

The stories move from enclosure to sensing, from sensing to intervention, and from intervention to autonomy. The destination is not total control. Living systems do not submit to that fantasy. The destination is a disciplined capacity to observe state, interpret drift, intervene at the least destructive point, verify the response, and learn.

Visual Program

Author's Note

Four Kinds of Claim

A book about a universal control layer can become irresponsible very quickly.

The phrase sounds as though a hidden master switch has been discovered. It has not. There is no single anatomical layer, molecule, organ, algorithm, or command center that controls a person, a habitat, or a planet. The term **Universal Control Layer** is the name I give to a proposed engineering architecture: the capability to measure a living system for a defined purpose, interpret changes in context, modulate the system at an appropriate point, verify the result, and update the model. It is a way of organizing the work that remains to be done.

Because the book moves between history, biology, engineering, medicine, and future spaceflight, the reader deserves to know which kind of statement is being made. I use four kinds of claim.

Documented event. These are historical facts supported by primary records or strong scholarly accounts: the decline of oxygen inside Biosphere 2; the 1976 outbreak associated with the Bellevue-Stratford Hotel; the creation of newborn screening for phenylketonuria; the NASA Twins Study; the decline in AIDS mortality after effective combination therapy; and the technical success and human conflict of Apollo 7. Historical interpretation can still differ, but the event itself is established.

Established science. These are findings supported by accepted bodies of evidence: living systems maintain themselves through feedback; microbial communities alter chemical environments; electrical signaling depends on ion gradients and molecular machinery; mass spectrometry detects many features that cannot yet be confidently identified; combination therapies can suppress adaptive pathogens more effectively than a single vulnerable mechanism; and physiological closed-loop devices require sensors, control logic, actuators, safeguards, and verification.

Engineering synthesis. This is the central contribution of the book. I connect established findings into an operating proposition: the same control questions recur in a patient, a greenhouse, a hospital, a spacecraft, a farm, and a planet. What state matters? How will it be sensed? What baseline is meaningful? What change is dangerous? What intervention is available? What unintended effects can the intervention produce? How will the system know whether it worked? The architecture is proposed, not proved as a universal law.

Future scenario. Some chapters look ahead to adaptive diagnostics, ISMM-informed monitoring, rational multi-target design, local biomanufacturing, and increasingly autonomous health and habitat management in deep space. These are not forecasts with guaranteed dates. They are scenarios built from capabilities that exist in partial form today. Their purpose is to expose the integration problem before the technology arrives in pieces.

The distinction matters because possibility is not proof. An unknown mass-spectrometry feature is not automatically a biomarker. A microbial molecule is not automatically a medicine. A multi-target profile is not automatically selective. An artificial-intelligence model is not automatically causal. A system that can intervene is not automatically safe.

The standard should rise as the consequence rises.

That sentence belongs equally to science and control engineering. A system may explore widely when the cost of error is low and reversible. It must become increasingly disciplined as its decisions approach a patient, a food supply, a community, or an autonomous spacecraft. The first duty of a control system is not to act. It is to know what it does not know.

This book therefore argues for ambition and restraint at the same time. We should be far more ambitious about measuring the chemical states that living systems use. We should be far more restrained about pretending that a partial measurement gives us permission to command the whole.

Preface

The Third Question

The first book in this trilogy asked a scientific question:

What are we missing?

The Ghost Genome examined the chemical output of living systems that cannot be read directly from one genome. Human cells, microbes, viruses, diet, environment, cellular turnover, local metals, oxygen, pH, and time participate in transformations whose products can be transient, dilute, spatially confined, or absent from reference libraries. The book's argument was not that every unknown molecule matters. It was that biology's functional chemistry is larger than the catalog we currently use to explain it.

The second book asked a managerial question:

How should we attack a problem when the map is incomplete?

Breakthrough Aggressiveness proposed that rigor and velocity are not rival religions. They are operating choices. A team must match the tempo, proof burden, risk posture, and tolerance for recoverable failure to the problem in front of it - and then change that posture when the problem changes. The central failure was mismatch: pushing too hard where consequence demands proof, or moving too slowly where the cost of delay is the greater danger.

This book asks the third question:

What do we build when the map becomes usable and the organization learns how to move?

The answer cannot be another catalog.

A catalog tells us what has been seen. A control system must tell us what is happening now, whether it matters, what can be changed, and whether the change helped. That difference separates observation from operation.

In manufacturing, I learned that measurement becomes valuable only when it closes a loop. A gauge that reports a dimension after a thousand defective parts have been made is an archive. A gauge connected to a process that can detect drift, identify the source, correct it, and verify the correction is a control system. The same principle appears in medical imaging. The instrument does not heal the patient, but it can turn an invisible condition into a decision. In environmental engineering, a sample is useful only if it represents the system well enough to guide action. In drug discovery, a binding score matters only if it survives the long translation into relevant exposure, effect, safety, manufacturability, and human benefit.

Living systems make this problem difficult because they do not hold still for inspection.

A cell changes when it is sampled. A microbial community changes when it is removed from its neighbors. Blood changes by the minute. The same molecule can act differently by dose, location, timing, partner, and chemical form. A habitat can look stable while its reserve capacity is being consumed. A person can remain within a population reference range while moving rapidly away from his or her own baseline. The output of one organism can become the input of another, then be edited by the host, then return to the community as a new condition.

The object under control is not a machine with fixed parts. It is a relationship that keeps remaking its parts.

This is why the phrase **biological software** is useful if we use it carefully. DNA, cells, organs, membranes, proteins, and structures give life durable machinery. Small molecules, ions, gradients, metabolites, peptides, gases, lipids, redox states, and signals carry much of the changing state that tells the machinery what to do next. Stop molecular movement and the electrical system also stops: neuronal and cardiac signaling depend on maintained ion gradients, channels, pumps, transmitters, energy, and reset. The electrical and chemical descriptions are not competitors. The electrical event is made possible by molecular conditions.

Software is not a perfect metaphor. Biological chemistry is not written by one programmer, and the organism does not execute a clean sequence of commands. The code changes while it runs. Signals are ambiguous. Components have histories. The system learns, adapts, compensates, and occasionally turns an intervention into the next problem.

But the metaphor makes one omission visible. Modern biology became exceptionally powerful at describing hardware and instructions. Medicine and engineering now need an equally serious capability to interpret the moving chemical state.

The Universal Control Layer is my name for that capability.

At its simplest, the architecture contains seven questions:

1. **Purpose:** What human or system outcome are we trying to protect or improve?
2. **State:** Which variables together describe the condition that matters?
3. **Sensing:** How can those variables be measured at the needed place, resolution, frequency, and confidence?
4. **Interpretation:** What change is normal adaptation, what is recoverable drift, and what is a causal warning?
5. **Modulation:** What is the least destructive intervention capable of changing the trajectory?
6. **Verification:** Did the intervention produce the intended effect without creating a worse state elsewhere?
7. **Learning:** How should the baseline, model, threshold, and intervention library change after the result?

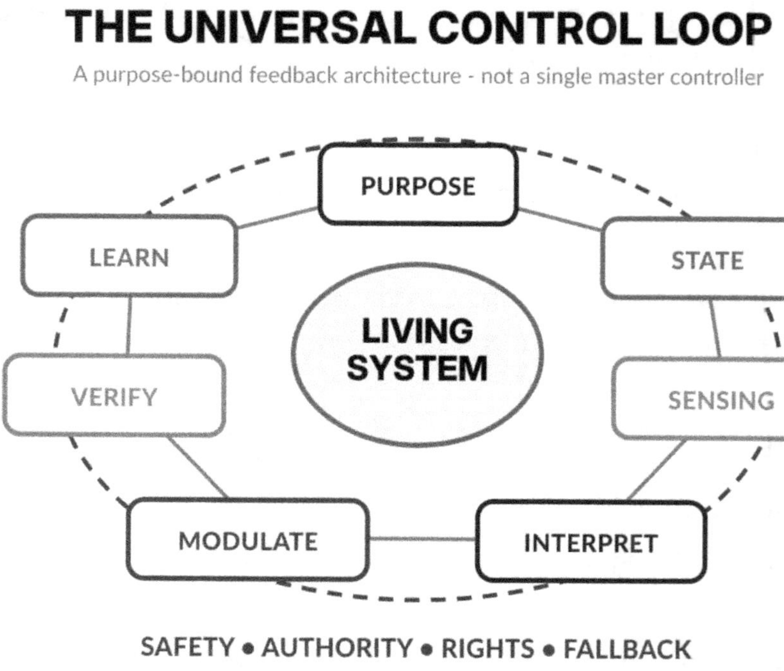

Figure 1. The Universal Control Loop. Purpose defines the state to be measured; sensing and interpretation guide an intervention; verification and learning update the model. Safety constraints surround the loop.

A thermostat is a closed-loop controller. It senses temperature, compares it with a set point, and turns a furnace or air conditioner on or off. A modern artificial pancreas is much more demanding: glucose sensing, insulin-delivery algorithms, delay, meals, exercise, sensor error, and individual physiology interact in a human whose safety cannot be reduced to one number. A habitat controller adds air, water, food, surfaces, waste, plants, microbes, crew behavior, energy, and mechanical equipment. Spaceship Earth adds oceans, soils, atmosphere, biodiversity, economies, politics, and billions of people who do not share one set point.

The larger the loop, the more dangerous the fantasy of a single controller.

The Universal Control Layer is therefore not a proposal for central command over biology. It is a proposal for distributed, purpose-bound, evidence-calibrated feedback. Different loops operate at different scales. A patient's bedside system should not govern a watershed. A farm's soil-monitoring system should not define a population's medical normal. A spacecraft may require more autonomy than a hospital because communication delay removes immediate ground control. The architecture is common; authority, evidence, and safeguards remain local to consequence.

The book begins with the largest sealed ecological experiment ever built. In September 1991, eight people entered a glass world in the Arizona desert. They had crops, animals, soil, water, air, and an extraordinary collection of sensors. They had designed a miniature planet.

Sixteen months later, the oxygen level had fallen to the equivalent of a high mountain.

No window was broken. No machine had obviously failed. The plants were still green.

The system had been answering a question the designers had not known to ask.

Measure for purpose.
Modulate for purpose.
Verify. Learn.

PART I

EVERY CLOSED LOOP IS ALIVE

CHAPTER 1

The Oxygen That Disappeared

A sealed world revealed that an inventory of parts is not a model of living state

---

On September 26, 1991, eight people walked toward a pair of air-lock doors in the Arizona desert and waved to the cameras.

Behind the glass waited a world.

Biosphere 2 covered a little more than three acres, but the number made the structure sound smaller than it felt. Steel frames held thousands of panes of glass over a miniature ocean, a marsh, a savanna, a desert, a rainforest, an intensive agricultural area, workshops, living quarters, and two enormous variable-volume chambers called lungs. The lungs rose and fell with heat and cooling so that changing air pressure would not tear the sealed structure apart.

The crew - Jane Poynter, Taber MacCallum, Abigail Alling, Mark Nelson, Roy Walford, Sally Silverstone, Mark Van Thillo, and Linda Leigh - stepped inside for a planned two-year closure. The experiment was audacious and theatrical because it had to be both. If human beings were ever going to live on the Moon, Mars, or another world, they would need more than a capsule stocked with supplies. They would need a small biosphere capable of recycling air, water, nutrients, and waste while growing food and preserving the people inside.

The doors closed.

For a while, the most visible problem was hunger. The agricultural system produced food, but not with the abundance the planners had hoped. The crew worked long days planting, harvesting, repairing, cleaning, measuring, and cooking. They became lean. Personal tensions rose. Injuries and management disputes gave reporters material for a story that increasingly sounded less like science and more like reality television before reality television had a name.

The more consequential problem was invisible.

Oxygen had begun near the ordinary atmospheric level of about 21 percent. It fell steadily. The decline was slow enough that a person waking each morning could not feel the difference from the day before. The body adapted until adaptation itself became part of the danger. Crew members tired more easily. Sleep became disturbed. Mental arithmetic took effort. Climbing stairs felt different. After roughly sixteen months, oxygen had fallen to about 14.5 percent - similar to the availability at an altitude of more than thirteen thousand feet.

The glass shell was remarkably tight. The missing oxygen had not simply leaked away.

If oxygen was being consumed by respiration, carbon dioxide should have risen in parallel. It did rise and fall with day, night, season, plant growth, and management, but not enough to balance the accounting. That made the mystery sharper. In a sealed system, matter does not vanish because the spreadsheet is inconvenient.

The answer turned out to be a chain.

The soils selected for Biosphere 2 contained substantial organic carbon. Microorganisms in those soils respired vigorously, consuming oxygen and releasing carbon dioxide. The carbon dioxide then reacted with exposed concrete, forming carbonates and removing part of the gas from the atmosphere. One process consumed the oxygen. Another hid much of the expected carbon-dioxide evidence.

No single component had behaved impossibly. Soil microbes respired. Concrete carbonated. Plants photosynthesized. People and animals consumed oxygen. The failure came from the interaction of ordinary processes at a system scale no one had calibrated well enough.

The oxygen did not disappear.

The model lost track of it.

The first mistake: confusing an inventory with a state

Biosphere 2 contained extraordinary knowledge. Its designers had selected species, estimated crop yields, engineered air handling, built water systems, modeled energy, and instrumented the enclosure. They knew how many people would live inside, what they would eat, how much carbon the plants might fix, how much oxygen people might consume, and how waste could be cycled back into production.

The parts list was sophisticated.

The state model was incomplete.

An inventory answers: What is present?

A state model answers: What is happening among what is present, at this moment, in this place, under these conditions, with what reserve capacity and what direction of travel?

The difference is the foundation of this book.

A patient can have a normal number of immune cells while those cells are entering an abnormal activation state. A farm can contain adequate nitrogen while the nitrogen is unavailable to a plant or being lost through microbial transformation. A reservoir can meet a chemical standard while a microbial community is shifting toward a bloom. A spacecraft can have acceptable humidity on average while condensation forms behind one panel and creates a local habitat. A company can have every required function on an organization chart while decision latency quietly consumes the opportunity.

Figure 2. Inventory Is Not State

An inventory of parts is not a model of living state.

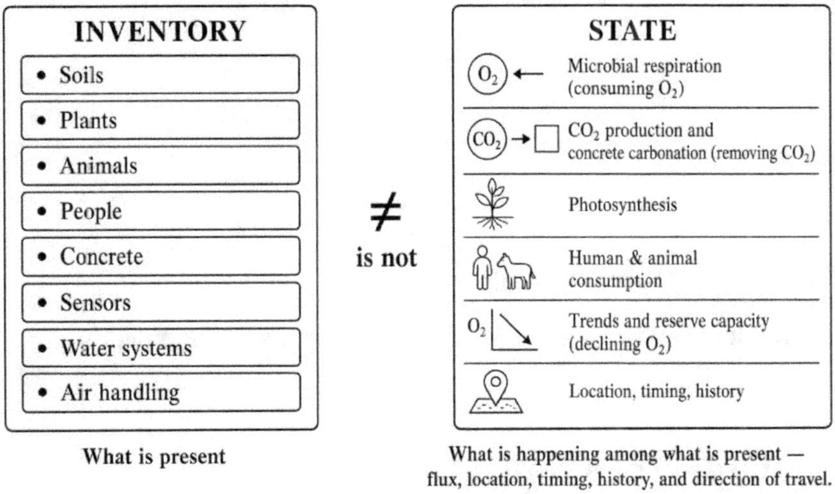

Figure 1. The Universal Control Loop. Purpose defines the state to be measured; sensing and interpretation guide an intervention; verification and learning update the model. Safety constraints surround the loop.

The inventory is not false. It is simply not alive enough.

Living state includes flux: what is being made, consumed, moved, transformed, stored, released, and sensed. It includes location. Ten milligrams of a compound in a bloodstream is not equivalent to ten milligrams concentrated at a tissue boundary. It includes timing. A pulse that lasts thirty seconds can trigger a durable response while remaining invisible to a weekly sample. It includes history. A system that has repeatedly recovered from stress may respond differently from one that has consumed its reserves.

The Ghost Genome described one reason the state is difficult to read. The relevant chemistry may be produced through relationships that do not belong to one genome: a microbial intermediate altered by another organism, conjugated by the host, exposed to dietary chemistry, bound to a metal, and degraded as quickly as it acts. A DNA sequence can reveal capacity. It cannot, by itself, reveal the complete chemical state produced by a living system at a particular moment.

This book begins where that map leaves off.

Once we accept that important state is hidden, what must a system do about it?

It must measure for purpose.

Then it must modulate for purpose.

And it must know whether its intervention improved the trajectory or merely moved the problem.

The anatomy of a control loop

Control engineering can sound remote from biology until we strip it to its essentials.

A controlled system has a desired outcome. It has a variable or set of variables that help describe whether the outcome is being achieved. It has sensors. It has a model or rule for interpreting the sensor data. It has an actuator capable of changing something. It has feedback that reveals the result. It has constraints that define what the controller must never do.

A thermostat uses room temperature as the measured variable and a furnace as the actuator. The model is crude: below the set point, add heat; above it, stop. Even this simple loop encounters delay, overshoot, sensor placement, open windows, sunlight, different rooms, and people who disagree about comfort.

Biology multiplies every difficulty.

The desired state is often a range rather than one number. The measured variables may be proxies. The intervention can take hours, days, or months to act. The system adapts to the intervention. The sensor can perturb the thing being measured. Two people with the same reading may require different responses. A short-term correction can create a long-term cost.

The U.S. Food and Drug Administration uses the term *physiologic closed-loop control* for medical devices that sense a physiological variable and automatically adjust the delivery or removal of energy or a substance. The definition is practical because it makes the components visible: sensor, control algorithm, actuator, patient, and feedback. It also makes the hazards visible. A bad sensor, wrong model, delayed response, actuator failure, or unusual patient condition can turn automation from protection into harm.

The Universal Control Layer extends the same architecture beyond one device or variable. It asks how a patient, habitat, farm, hospital, or spacecraft can develop a layered capability to sense and respond to biological state.

The word *universal* refers to the recurrence of the architecture, not to one universal controller.

Every serious loop needs seven elements.

Purpose. The loop must protect or improve something specific. "Optimize health" is too vague. Maintain glucose within a safe range without dangerous hypoglycemia is more useful. Preserve crew cognitive function during a six-month transit is more useful. Prevent a water system from entering a Legionella-amplifying state is more useful.

State. The loop must identify the smallest set of variables that together describe the condition relevant to the purpose. A single variable is attractive because it is easy to display. It is often insufficient.

Sensing. The measurement must occur at the right location, resolution, and cadence. An accurate monthly measurement can be useless when the dangerous transition occurs in hours.

Interpretation. The system needs a way to distinguish expected variation from meaningful drift. That requires context, baselines, confidence levels, and causal humility.

Modulation. The loop needs an intervention that can influence the state. The best intervention is not necessarily the most powerful. It is the least disruptive action capable of changing the trajectory with adequate confidence.

Verification. A control action without a verified response is not control. It is hope with machinery attached.

Learning. The system must update its model after every meaningful result. Otherwise it repeats yesterday's assumptions with greater automation.

Surrounding the loop are safeguards: human override, fail-safe states, bounded authority, independent validation, security, ethics, and explicit conditions under which the controller must stop acting.

Biosphere 2 had many sensors and many actuators. Crew members could plant faster-growing crops, irrigate or dry areas, remove biomass, operate carbon-dioxide scrubbers, inject oxygen, and change routines. What it lacked at the beginning was a sufficiently complete model connecting the measured signals to total system metabolism.

That is not an indictment of the experiment. It is the reason the experiment mattered.

A sealed system turns hidden assumptions into measurable consequences.

Earth is closed - and not closed

“Spaceship Earth” is one of those phrases that can illuminate and mislead in the same sentence.

Earth is nearly closed with respect to matter. We receive meteorites and lose some atmospheric material, but for practical human timescales, the atoms available to civilization are overwhelmingly the atoms already here. Waste does not go away. It moves. Carbon enters air, ocean, biomass, soil, concrete, or rock. Nitrogen changes form. Water changes location and quality. Plastics fragment. Metals disperse. A pollutant emitted in one place can appear in another population, another species, or another generation.

Earth is open with respect to energy. Sunlight arrives continuously and heat leaves. That flow drives climate, photosynthesis, circulation, and much of the planet's capacity to remain far from equilibrium.

A spacecraft habitat is more tightly closed in matter and much smaller in reserves. It may receive solar or nuclear energy, but resupply is limited, waste volume is unforgiving, and the distance to help grows. A human body is open to food, air, water, microbes, medicines, and information, yet it also maintains internal boundaries and recycles much of its material. These systems are not identical. They share the problem of finite inventories governed by moving biological processes.

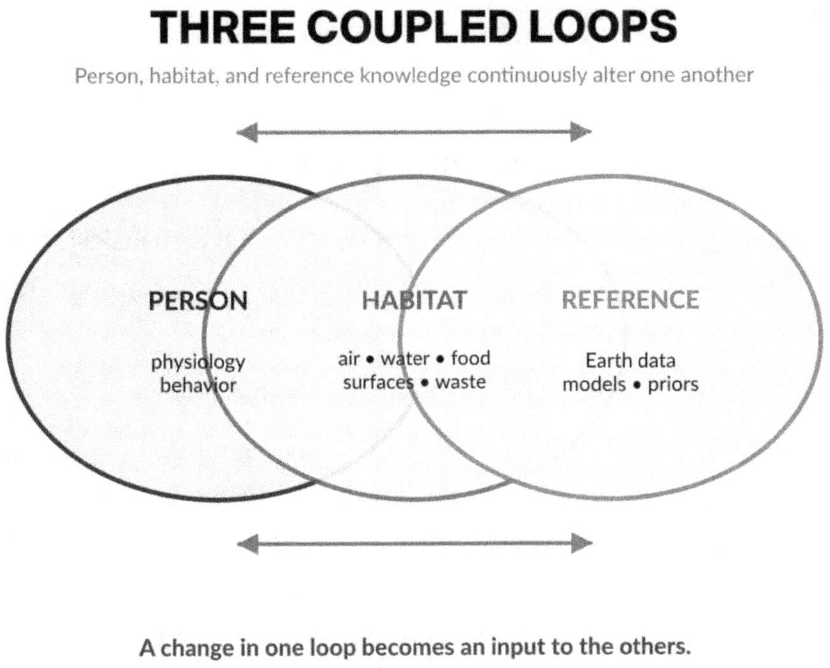

Figure 3. Three closed loops. A person, a habitat, and a planet operate at different scales and with different authorities, but each depends on sensing state, preserving reserves, managing flows, and learning from feedback.

The temptation is to treat the smallest loop as a miniature version of the largest. Biosphere 2 warned against that. Scale changes behavior. Surface-to-volume ratios change. Species diversity changes. reserve capacity changes. Human choice changes. Political legitimacy changes. A treatment appropriate for a patient cannot simply be enlarged into environmental policy.

What transfers is the discipline of the questions.

What purpose is being protected?

What state variables matter?

What is the baseline?

What is moving?

What can be changed?

Who has authority to change it?

What is the safe fallback if the model is wrong?

When the set point becomes the problem

The easiest control problem has a known set point. Oxygen inside a habitat must remain within a safe range. Blood glucose has a clinically useful target range. Water pressure must not fall below a threshold.

Complex living systems often have no single ideal number.

Consider body temperature. A fever can be dangerous, but it can also be part of an adaptive immune response. Automatically driving every elevation to one fixed value could interfere with the biology the system is trying to protect. Inflammation can damage tissue, but suppressing it indiscriminately can impair defense and repair. Microbial diversity can be associated with resilience in one habitat, but diversity alone does not distinguish harmless organisms from a dangerous contaminant.

A control system can therefore fail by optimizing the wrong variable perfectly.

This is the biological version of Goodhart's law: when a measure becomes a target, it can stop being a good measure. A hospital can reduce average length of stay while increasing readmissions. A farm can maximize annual yield while consuming soil structure and future water capacity. A drug can normalize one biomarker while failing to improve how a patient feels or functions. A spacecraft can meet a surface-cleanliness count while a biofilm grows inside a water line.

The Universal Control Layer begins with purpose because purpose constrains the metric. The metric is not the mission.

Biosphere 2's mission was not to keep a dashboard beautiful. It was to sustain eight people and a functioning ecology inside material closure. When oxygen fell, adding oxygen was not "cheating" in any scientifically meaningful sense if the purpose included learning how to manage a life-supporting system. The injection did change the purity of the closure experiment, and it had to be disclosed and interpreted. But the deeper mistake would have been sacrificing people to preserve a public narrative.

The control system serves the living purpose, not the theater of perfect control.

The hidden denominator: reserve capacity

A system can look stable because it is compensating.

The human body maintains blood pH within a narrow range through lungs, kidneys, buffers, and cellular processes. A normal pH does not prove the absence of stress. It may mean the compensatory machinery is working harder. A power grid can maintain frequency while reserve generation is disappearing. A team can meet milestones while its most capable people are absorbing unsustainable load. A habitat can maintain carbon dioxide through a scrubber while the biological imbalance producing the gas grows worse.

The measured value is the numerator. Reserve capacity is the hidden denominator.

This is why trends, perturbations, and recovery matter. A healthy system is not one that never changes. It is one that can change, absorb disturbance, and return without consuming an unacceptable portion of its future options.

In control language, resilience is not static sameness. It is the ability to remain within safe bounds, recover from shocks, and avoid being trapped in a degraded state.

For a human, reserve can include metabolic flexibility, immune competence, kidney function, cardiovascular capacity, microbial functions, sleep, nutrition, and social support. For a habitat, it can include oxygen margin, water storage, spare filters, alternative crops, microbial redundancy, energy reserves, and manual operating modes. For Earth, it includes biodiversity, soil carbon, freshwater, forests, climate stability, institutional trust, and the capacity of communities to adapt without collapse.

A useful Universal Control Layer must therefore monitor not only the target variable but the cost of holding it there.

How much insulin is required to maintain the glucose range?

How much energy is required to maintain the habitat temperature?

How much antibiotic pressure is required to suppress the pathogen?

How much irrigation is required to sustain the yield?

How much overtime is required to maintain the schedule?

When the control effort rises while the output remains constant, the system is telling us that its state has changed.

The danger of a universal dashboard

A single integrated dashboard is visually irresistible. It promises that complexity can be brought under one pane of glass. Green means safe. Yellow means attention. Red means act.

The ambition is useful. The compression can be dangerous.

Different signals have different confidence. Some are measured directly. Some are inferred. Some are delayed. Some are population averages. Some are personal baselines. Some reflect cause; others reflect consequence. Combining them into one score can create an authority the evidence does not deserve.

A rigorous control layer needs visible uncertainty.

It should distinguish:

- a directly measured variable from a model-derived estimate;
- an identified molecule from an unknown analytical feature;
- a correlation from a demonstrated causal link;
- a validated intervention from an experimental option;
- a local signal from a system-wide conclusion;
- a reversible action from an irreversible one.

The system should be allowed to say: *we do not know what this means yet.*

That is not a failure of intelligence. It is a safety function.

The most dangerous controller is not the one with missing data. It is the one that converts missing data into false certainty and acts at scale.

From Biosphere 2 to the hospital room

The control problem revealed in the desert already exists in medicine.

A clinician looking at a patient sees a partial dashboard: vital signs, laboratory values, imaging, symptoms, history, medications, and physical examination. Each is a window. None is the person. The values arrive at different times and resolutions. A blood test mixes chemistry from many tissues. A stool sample overrepresents what reaches the end of the intestine. A wearable measures what its sensors can reach. The most important process may be local, transient, or upstream of the measured consequence.

Modern medicine is nevertheless moving toward closed loops. Continuous glucose monitors guide insulin pumps. Ventilators respond to measured conditions. Pacemakers and neurostimulators can adapt to physiological signals. Intensive-care systems integrate streams of data to identify deterioration.

The next step is not simply more sensors. It is better state interpretation.

A million measurements can make the problem worse if the reference language is weak. An AI system trained on incomplete labels can become exceptionally fast at reproducing incomplete interpretation. A mass spectrometer can detect thousands of features, but a signal without structural identity, source context, temporal behavior, or biological relevance remains a clue rather than a command.

This is where the first two books in the trilogy enter without needing to be repeated.

The Ghost Genome supplies the scientific warning: the chemical map is incomplete.

Breakthrough Aggressiveness supplies the operating warning: the method must change as uncertainty, consequence, and maturity change.

The Universal Control Layer supplies the engineering question: how do we convert an expanding map and an adaptive organization into safe, useful feedback?

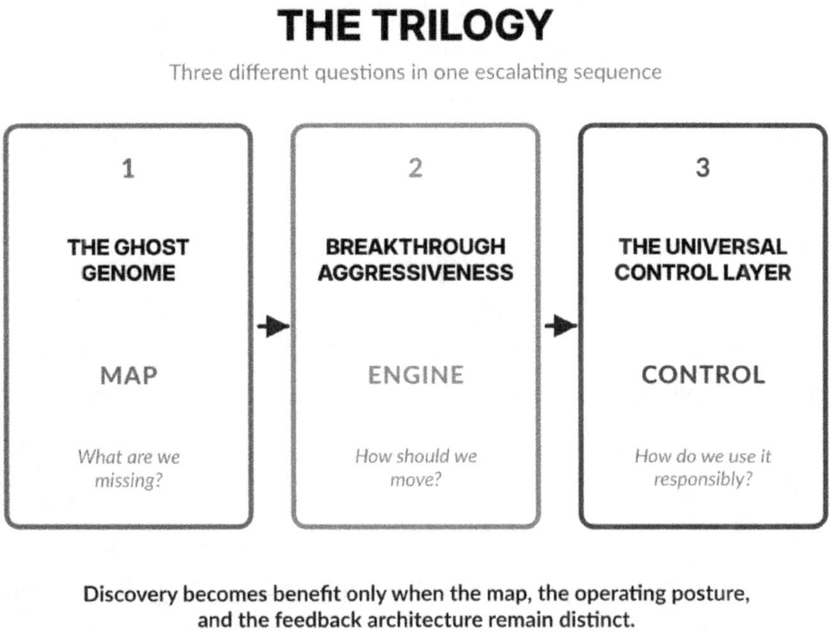

Figure 4. The trilogy. The Ghost Genome asks what chemistry is missing. Breakthrough Aggressiveness asks how to match the operating method to the problem. The Universal Control Layer asks how to sense, interpret, modulate, verify, and learn at system scale.

What the crew could feel before the model could explain

The oxygen decline was not experienced as a clean graph.

It arrived inside ordinary work.

A task that had been manageable became tiring. A calculation took longer. A climb required a pause. Sleep changed. Hunger and interpersonal strain complicated interpretation. The crew was both the population at risk and part of the sensor network. Their subjective experience mattered, but it was not specific. Fatigue could come from caloric restriction, workload, low oxygen, poor sleep, conflict, or several at once.

This is a recurring problem in human control systems. Symptoms are meaningful and ambiguous. A person often detects that something is wrong before an instrument identifies the state, but the person may not know the cause. The correct response is neither to dismiss the experience as subjective nor to let it substitute for measurement.

A mature loop uses both.

The crew report becomes a trigger for additional sensing. The sensing becomes a hypothesis. The hypothesis becomes an intervention. The intervention is verified against both measured state and lived function.

The same logic belongs in a hospital. A patient who says, "I am not right," may be reporting an early state not yet visible in routine values. In a spacecraft, a crew member's complaint can be an environmental sensor. In a factory, an operator may hear or feel a change before a vibration monitor crosses a threshold. In an ecosystem, a fisher or farmer can perceive a shift before a formal survey.

Local knowledge is not the opposite of data. It is one source of data with its own error structure.

Four ways a control system can be wrong

The oxygen mystery also reveals four distinct failure modes.

Model error. The sensors are working, but the relationships among variables are wrong or incomplete. Biosphere 2 underestimated the integrated effect of soil respiration and concrete carbonation.

Sensor error. The system state is understood in principle, but measurements are biased, delayed, misplaced, or unavailable.

Actuator error. The system knows what is wrong but cannot change it reliably. A scrubber may be undersized, a drug may not reach the tissue, or a valve may fail.

Governance error. The system has information and capability, but authority, incentives, communication, or public narrative prevents appropriate action.

These failures require different remedies. More data will not repair a bad actuator. A stronger actuator will not repair a wrong model. A correct model will not help if the organization suppresses the warning.

The Universal Control Layer should label the uncertainty by layer. "Low confidence" is too vague. Is confidence low because the signal may be false, because identity is unknown, because causality is uncertain, because the intervention is untested, or because authority is contested?

The label tells the operator what kind of learning is needed.

Learning without turning failure into theater

Biosphere 2 became a cultural argument. Supporters described a landmark experiment. Critics described secrecy, spectacle, and failure. The oxygen injection became symbolic: to some, it proved the project had broken its promise; to others, it was an appropriate life-support intervention inside a learning system.

The dispute matters because public science often confuses demonstration with experiment.

A demonstration is expected to work.

An experiment is expected to reveal what was not known.

Biosphere 2 was sold with some of the confidence of a demonstration and operated with the uncertainty of an experiment. When the system produced surprises, the public judged the surprise as failure of competence rather than output of inquiry.

A future control-layer program should define which mode it is in.

During discovery, the objective may be to expose hidden dynamics. During qualification, the objective is to prove repeatable performance.

During mission operation, the objective is to keep people safe. The same anomaly has different meaning in each mode.

The organization must migrate its proof standard and communication accordingly. Otherwise it will either hide useful failure or expose people to an experiment they believed was a service.

Control as a system of systems

The Universal Control Layer will never be one device.

It will be a system of systems: sensors produced by different vendors, models maintained by different teams, interventions regulated through different authorities, and human decisions made under different incentives. The interfaces become as important as the components.

One system may report oxygen as partial pressure, another as percentage, another as physiological effect. One database may identify a molecule at high confidence, another only a feature family. One model may update daily, another annually. One actuator may respond in seconds, another in weeks.

The control architecture needs translation rules, common time stamps, version control, provenance, and explicit ownership of handoffs.

A failure at an interface can look like a biological surprise.

The first step toward universality is not one universal platform. It is interoperability around purpose and evidence.

The first law of the Universal Control Layer

The first law is simple:

> ***Inventory is not state.***

A list of organisms is not the chemistry they are producing.

A list of genes is not the pathway activity now.

A list of drugs is not the exposure in the relevant tissue.

A list of crops is not the oxygen balance of a biosphere.

A list of sensors is not situational awareness.

A list of safety procedures is not safety culture.

State is relationship in motion.

That sentence changes the way a system should be designed. Instead of asking only what components must be present, we ask what flows must be observed. Instead of assuming equilibrium, we monitor direction and reserve. Instead of treating unexplained signals as disposable noise, we preserve them with confidence labels. Instead of choosing the most forceful intervention, we choose the smallest intervention capable of restoring the trajectory. Instead of trusting the model because it is complex, we test whether it predicts what happens next.

Biosphere 2 did not fail to teach us how to live in space. It succeeded in revealing why living in space will be harder than sealing people beside plants.

The experiment exposed the true unit of life support: not the species list, the machine list, or the oxygen tank, but the coupled metabolism of the whole.

And it revealed the moral center of control engineering.

When the model and the living system disagree, the living system gets the final word.

The next chapter begins with a man who had mastered the factory so completely that he believed its logic could be moved into the Amazon. He planted millions of rubber trees in orderly rows.

The rows were exactly what the enemy needed.

CHAPTER 2

The Plantation That Invited Its Enemy

Industrial order removed the ecological spacing that had been doing invisible work

---

In 1928, two ships moved up the Tapajos River carrying the pieces of an American future.

The *Lake Ormoc* and the *Lake Farge* brought machine tools, trucks, tractors, construction equipment, prefabricated buildings, hospital supplies, furniture, plumbing, and the components of a water tower. Henry Ford had acquired an enormous concession in the Brazilian Amazon. His company needed rubber for tires, belts, hoses, gaskets, and the growing appetite of mass production. British and Dutch interests dominated much of the world supply. Ford intended to control the material at its source.

Vertical integration had already helped make him one of the most powerful industrialists on Earth. Iron ore, timber, coal, railroads, glass, assembly, shipping: the River Rouge complex near Detroit had turned a supply chain into a single coordinated machine. Raw material entered one end. Finished automobiles emerged from the other.

The Amazon was going to become another Rouge.

The settlement was named Fordlandia. American-style houses rose above the river. There was a hospital, school, sawmill, power plant, water system, recreation hall, swimming pool, and streets laid out with the confidence of a company town that expected the jungle to recognize a superior plan. Brazilian workers were introduced to time clocks, cafeteria food, housing inspections, alcohol restrictions, and ideas about health and discipline imported from Michigan.

The rubber trees were planted in rows.

That decision looked like efficiency. It was also the beginning of the end.

In the Amazon forest, *Hevea brasiliensis* does not ordinarily grow as a dense plantation. The trees are dispersed among other species. The spacing makes tapping less convenient, but it also interrupts the movement of specialized pests and pathogens. Distance, diversity, and forest structure perform protective work.

Ford's plantation removed that work.

Millions of genetically similar hosts stood close enough for South American leaf blight and other pests to move through them with industrial efficiency. The plantation did not merely fail to reproduce the forest. It reorganized the environment on behalf of the enemy.

The trees suffered. The management culture suffered too. In December 1930, workers rebelled against cafeteria rules and other imposed conditions in an uprising remembered as *Quebra-Panelas* - the breaking of the pans. Managers fled or took shelter while parts of the settlement were damaged. The project shifted to Belterra, where terrain and agricultural conditions were better, but the dream of a Ford-controlled Amazon rubber supply never became what its founder intended. In 1945, the company transferred the properties back to the Brazilian government at a large loss.

Not one drop of Fordlandia rubber became a Ford tire.

The common lesson is that Henry Ford did not understand the Amazon.

The deeper lesson is that he understood his own system too well.

The success trap

A successful operating model becomes more than a method. It becomes a theory of reality.

Ford's factories had demonstrated that variability could be reduced, tasks could be standardized, parts could be made interchangeable, and flow could be accelerated through coordination. In the right environment, these principles transformed manufacturing and consumer life.

The plantation carried those principles into a system where variability was not merely waste. Some of it was protection.

This distinction appears everywhere biological systems meet industrial logic.

A factory often seeks uniform inputs because uniformity makes output predictable. An ecosystem frequently uses heterogeneity to prevent one disturbance from reaching everything at once. A factory creates short paths between compatible steps. An ecosystem may use distance to interrupt contagion. A factory removes redundancy to lower cost. A living system may retain apparently inefficient redundancy because different components fail under different stresses.

The industrial engineer asks: What can be simplified without changing the product?

The systems biologist must add: What function disappears when the complexity disappears?

Fordlandia answered after the rows were planted.

The spacing of the wild rubber trees had not looked like a technology. It had no inventor, patent, maintenance manual, or control panel. Yet it performed a control function: it lowered transmission efficiency for an enemy adapted to that host.

The forest architecture was part of the plant's immune environment.

When the architecture was removed, the pathogen did not need to become more aggressive. The system became easier to attack.

The second law: do not simplify away the regulator

The second law of the Universal Control Layer is:

> ***Do not simplify away the function you have not yet learned to measure.***

This law is not an argument against simplification. Science depends on controlled reduction. Engineering depends on models. Manufacturing depends on standardization. Medicine depends on isolating variables. The question is whether the removed context was noise or control.

That cannot always be known in advance. It can be tested.

When a microbial community is reduced to one cultured species, does the chemical output remain the same?

When a tissue is dissociated into cells, does the signaling behavior remain?

When a farm is converted from a diverse rotation to one crop, what happens to pest pressure, soil structure, nutrient cycling, and resilience?

When a drug is optimized for one target, what compensatory pathways become dominant?

When a hospital water system is made more energy efficient, do temperature and stagnation patterns create a microbial niche?

When a spacecraft is designed around nominal crew metabolism, what happens during illness, stress, dietary change, or microbial drift?

The experiment should be designed not only to find the average effect but to reveal the lost function.

A control layer therefore needs a record of what was removed.

This sounds obvious. It is rarely practiced. Teams document the variable they choose to study and treat the discarded context as background. Later, when the result fails to translate, the missing context is rediscovered as a confounder.

The word *confounder* can hide an important truth: sometimes the confounder is the system.

Monoculture is a control decision

Agricultural monoculture is not inherently irrational. Uniform crops simplify planting, harvesting, processing, quality control, and markets. Large-scale production has helped feed billions of people. The danger comes when operational efficiency is mistaken for biological security.

The Irish potato famine is often reduced to dependence on one crop. The more exact vulnerability involved a narrow genetic base and a pathogen capable of exploiting it. The twentieth-century spread of Panama disease through the global Gros Michel banana industry revealed a similar fragility. The 1970 southern corn leaf blight epidemic in the United States exploited widespread use of a particular cytoplasmic trait. In each case, efficiency had concentrated susceptibility.

The lesson is not diversity at any cost. Diversity can reduce yield, complicate logistics, and introduce its own hazards. The lesson is that diversity has a control value that must be priced rather than romanticized.

A Universal Control Layer for agriculture would not ask only, "How many species are present?" It would ask which functions are redundant, which are concentrated, which pathways share a common failure mode, and how rapidly a pathogen or environmental shock could propagate.

That requires a different state vector.

A **state vector** is a set of variables that together describe a system well enough for the purpose of control. The concept comes from engineering and mathematics, but the discipline is broadly useful. No single variable tells us whether a rubber plantation is resilient. Relevant variables might include host genetics, canopy structure, soil moisture, pathogen pressure, insect populations, nutrient status, spatial connectivity, weather, and the response of neighboring species.

The vector does not need to include everything. It needs to include enough to distinguish the states that require different action.

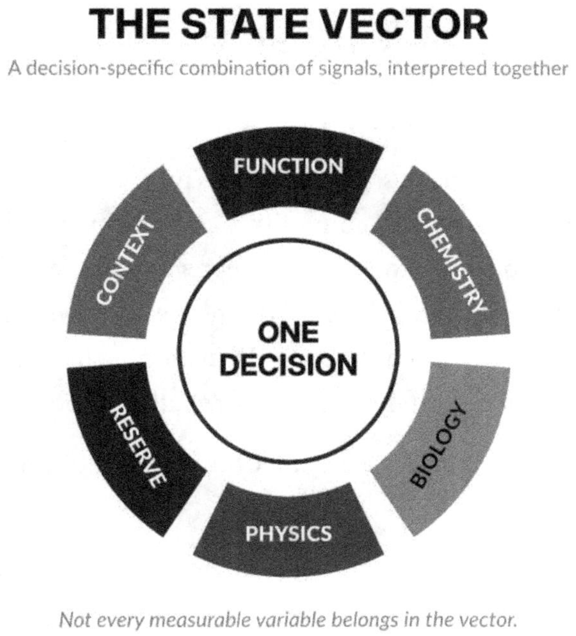

Figure 5. The state vector. A useful control picture combines identity, abundance, location, flux, time, reserve capacity, and confidence. The exact variables depend on the purpose.

This is where many "digital twin" projects become less useful than their name suggests. A digital twin is not valuable because it contains a beautiful representation of a physical system. It is valuable if it preserves the variables and relationships needed to predict how the system will respond to a meaningful intervention.

A perfect three-dimensional model of every rubber tree would not have saved Fordlandia if it omitted pathogen transmission.

A hospital digital twin that models bed occupancy but not water-system temperatures can miss the next outbreak.

A patient twin that includes genomic variants but not exposure, medications, diet, microbial metabolism, or time can be precise about the wrong state.

The twin must contain the enemy's path.

The invisible services of context

Nature performs services without sending invoices.

A wetland slows water, traps sediment, cycles nutrients, and supports organisms. A diverse microbial community occupies niches that might otherwise be available to an invader. Soil aggregates regulate air and water. Mucus creates a physical and chemical boundary. The liver modifies compounds before they reach systemic circulation. Behavioral norms can interrupt transmission. Distance can be a firewall.

Because these services are distributed, they become visible mainly when they fail.

Fordlandia's spacing problem offers a clean example, but the same logic extends to human health.

The gut microbiome is often described as a list of organisms. Its control functions may include competition for resources, production of inhibitory compounds, transformation of bile acids, maintenance of barrier conditions, and influence on immune development. Removing organisms with a broad-spectrum antibiotic can be necessary and lifesaving. It can also remove colonization resistance and create an opening for *Clostridioides difficile*. The antibiotic did not "cause" the pathogen in a simple sense. It changed the ecological control environment.

In cancer, a tumor is not only a collection of malignant cells. It is also vasculature, immune cells, fibroblasts, extracellular matrix, metabolites, oxygen gradients, and mechanical forces. A therapy that kills one cellular population can select for another. The intervention changes the environment in which the next state will emerge.

In organizations, informal relationships can perform the same invisible function. A veteran engineer may connect teams, translate risk, and stop problems before they enter formal systems. Removing the role because it does not appear on a process map can increase apparent efficiency while destroying coordination.

The recurring error is to value only the component whose output is easily counted.

Control begins by identifying the quiet functions that keep disturbances local.

Diversity, redundancy, and the price of preparedness

Preparedness always looks inefficient before the event.

A spare pump sits unused. A second supplier costs more. A genetically diverse seed bank does not maximize this year's yield. Manual operating capability requires training for a mode everyone hopes never to use. A library of natural compounds may contain thousands of inactive fractions before one reveals a useful function.

The economic argument against redundancy is often persuasive because the benefit is probabilistic and the cost is immediate.

The Universal Control Layer does not solve this tension. It makes it measurable.

Three questions matter.

What failure does the redundancy protect against?

A generic appeal to resilience can justify anything. The threat should be specified: one pathogen, one supplier, one sensor failure, one target escape pathway, one power interruption.

How correlated are the supposed alternatives?

Two suppliers in the same region may fail together. Two antibiotics with the same resistance mechanism are not meaningful redundancy. Two sensors built on the same physical principle may share a blind spot.

How much warning will the system provide before the reserve is needed?

If warning is long, capacity can be mobilized. If warning is short, reserve must already exist.

Fordlandia had millions of trees but little meaningful biological redundancy. They were connected by shared susceptibility and spatial arrangement.

Quantity created the appearance of reserve. Correlation removed it.

The control layer as a portfolio

One of the practical implications is that control should be designed as a portfolio rather than a single line of defense.

A biological portfolio can include prevention, early detection, containment, treatment, and recovery. A habitat portfolio can include passive design, active sensing, automated response, manual override, spare capacity, and the ability to isolate a subsystem. A therapeutic portfolio can include a highly selective agent, a rational combination, a local intervention, and a fallback mechanism if resistance emerges.

The components should not all fail for the same reason.

This sounds like ordinary risk management. The difference is that the Universal Control Layer treats hidden chemistry and biological adaptation as dynamic participants. The threat is not static, and the intervention changes the threat.

A fungicide alters selection. A disinfectant changes the surface community. A drug changes the metabolic environment. A diet changes substrate availability. A habitat filtration system changes humidity and particle distribution. A policy changes behavior.

The controller must therefore monitor the response it creates.

An intervention library without response monitoring is a menu, not a control system.

The temptation to declare nature optimized

Fordlandia can produce the opposite romantic error: the belief that natural systems are perfectly optimized and industrial intervention is inherently foolish.

Nature is not optimized for human goals. Evolution works with history, trade-offs, local fitness, and available variation. Natural systems contain disease, predation, waste, cruelty, collapse, and extinction. Rubber-tree spacing helps manage leaf blight in the forest, but it does not produce an efficient supply of tires. A natural molecule can be a toxin. A stable ecosystem can be inhospitable to people.

The purpose of reading natural control is not to surrender engineering judgment. It is to avoid erasing accumulated function before understanding it.

Industrial systems bring advantages nature does not provide: reproducibility, sanitation, scale, standardization, quality assurance, and the ability to define a human purpose. Biological systems bring strategies industrial models often miss: distributed sensing, feedback, redundancy, adaptive response, compartmentalization, and chemistry shaped by long interaction.

The strongest architecture combines them.

This is why I resist the false choice between synthetic chemistry and natural products. Synthetic chemistry is an extraordinary means of making, modifying, and optimizing matter. The problem arises when the design space is detached from biological priors and the resulting throughput is mistaken for relevance. Nature is not the only designer. It is a source of physically realized hypotheses.

The Universal Control Layer needs both the natural archive and the engineered actuator.

We measure what living systems already do. We identify which functions matter. We create interventions that reproduce, amplify, attenuate, redirect, or replace those functions. Then we verify in the system that matters.

A practical test: the context-removal audit

Before simplifying a living system, ask five questions.

1. **What context is being removed?** Species, spatial structure, extracellular matrix, host tissue, diet, metabolites, timing, physical gradients, behavior, or environmental exposure.
2. **What control functions might that context perform?** Competition, buffering, transport, sequestration, signaling, repair, containment, or error correction.
3. **How will we know if the function has disappeared?** Define an observable response before the reduction.
4. **Can the context be removed in stages?** Progressive reduction often reveals where function collapses.
5. **What result would require restoring complexity?** Precommit to the condition rather than explaining failure after the fact.

This audit belongs in cell biology, assay design, agriculture, habitat engineering, organizational redesign, and AI model development. A training dataset is also a simplified environment. If it excludes negative results, unusual chemotypes, transient states, spatial data, and failed interventions, the model learns a Fordlandia version of biology: neat rows of what the archive was built to retain.

The state of the enemy

The most powerful line in a control system may be the one describing the adversary's opportunity.

For leaf blight, the opportunity was dense access to susceptible hosts.

For antibiotic resistance, it may be prolonged exposure to a concentration that suppresses sensitive organisms without eliminating the resistant population.

For a hospital-acquired infection, it may be a warm, stagnant water line, aerosolization, and vulnerable patients.

For cancer, it may be a protected niche and compensatory signaling.

For misinformation, it may be a network structure that rewards rapid emotional replication.

For bureaucratic delay, it may be a review system in which no individual bears the cost of waiting.

The enemy is not always an organism. It is any process capable of using the system's structure against its purpose.

A Universal Control Layer should model opportunity, not only presence.

Detecting one pathogen cell may matter less than detecting the environmental state in which that organism can amplify. Detecting one resistant clone may matter less than measuring the selection pressure that will favor it. Detecting one toxic metabolite may matter less than recognizing the pathway that is accelerating its production.

Control becomes more effective as it moves upstream from the visible failure to the conditions that make the failure likely.

Fordlandia attacked the fungus after creating the plantation it wanted.

A better design would have treated spacing, genetic diversity, climate, soil, and pathogen ecology as part of production engineering from the beginning.

The revolt as a control signal

The worker uprising at Fordlandia is often treated as a colorful side story: Brazilians rejected bland cafeteria food and rigid American rules, smashed dishes and equipment, and drove managers into temporary retreat.

In systems terms, the revolt was feedback.

Management had specified not only production but diet, schedule, housing, alcohol, inspection, and social behavior. The company treated resistance as disorder introduced by the workforce. The workforce was reporting that the operating model did not fit the human environment.

A controller can respond to resistance in two ways.

It can increase force.

Or it can ask whether the resistance reveals model error.

This does not mean every complaint is correct or every local practice should override safety and quality. It means persistent friction is information about the fit between design and context.

Organizations routinely suppress this signal. A frontline team develops workarounds because the official process is slow. Management calls the workaround noncompliance. A patient stops taking a medicine because the regimen is intolerable. The system calls the patient nonadherent. A community resists an environmental intervention because risks and benefits are distributed unfairly. The program calls the community misinformed.

Sometimes the label is accurate. Sometimes it hides the failure to design an operable system.

The Universal Control Layer needs a channel for human feedback that is not filtered entirely through the authority that designed the intervention.

Local knowledge and the missing sensor

Ford's managers arrived with industrial expertise but little experience in tropical botany, soils, disease ecology, or the social organization of the region. The project did consult specialists at different stages, yet the governing model remained imported.

Local knowledge is often difficult to formalize. It can be anecdotal, tacit, inconsistent, or bound to conditions that have never been measured. That makes it easy for a technically trained team to discount.

The right response is translation, not worship or rejection.

What recurring observation does the local practitioner report?

Under what conditions does it hold?

What mechanism might explain it?

How can it be tested without destroying the context that makes it true?

Which part is cultural preference, and which part is hidden system knowledge?

A rubber tapper may not describe pathogen transmission mathematically. The pattern of successful trees across the forest can still contain an epidemiological result.

The control layer should make room for evidence that arrives before formal language.

The speed of the enemy

Industrial agriculture often gains efficiency by synchronizing the crop. Planting, growth, flowering, and harvest become more uniform. A pathogen can benefit from the same synchronization.

The speed of spread depends on connectivity, susceptibility, generation time, climate, and detection. A plantation can be optimized for the harvest cycle while leaving the pathogen with a much shorter iteration cycle.

This is the biological version of an asymmetry seen in cybersecurity and warfare: the attacker needs one usable path; the defender must protect many.

A control system should therefore compare cycle times.

How quickly can the threat adapt or move?

How quickly can the system detect it?

How quickly can the organization decide?

How quickly can the intervention be deployed?

If the combined detection-and-response time exceeds the threat's propagation time, the system is structurally behind even when every team meets its internal schedule.

Fordlandia measured work by the cadence of planting and construction. The leaf pathogen measured opportunity leaf by leaf.

The dataset plantation

The same simplification can occur in artificial intelligence.

A training dataset may be enormous and still be a monoculture. Records can share the same instruments, populations, publication biases, chemical libraries, and institutional assumptions. A model trained on billions of examples can become extremely capable within a narrow distribution and brittle outside it.

Scale is not diversity.

A chemical dataset dominated by easily synthesized, stable, rule-compliant structures can underrepresent macrocycles, transient metabolites, stereochemical complexity, negative results, and context-dependent natural products. A medical dataset can overrepresent people with access to particular health systems. An environmental dataset can sample accessible locations and miss the boundaries where important chemistry occurs.

The control-layer equivalent of genetic diversity is **distributional diversity**: different sources, conditions, methods, populations, outcomes, and failure modes.

The model should know which rows of its plantation share the same susceptibility.

The third system enters the story

By the middle of the twentieth century, architects and engineers were learning to seal buildings more tightly, circulate air more efficiently, and create indoor environments larger and more complex than earlier generations had imagined. Hotels, hospitals, convention centers, offices, and factories became mechanical habitats.

The people inside experienced conditioned air and running water as services. The systems behind the walls experienced temperature, flow, stagnation, aerosols, scale, corrosion, nutrients, and biofilms.

In July 1976, thousands of members of the American Legion gathered in Philadelphia. Within days, some began to develop fever, cough, chest pain, and pneumonia. The cause was unknown. The country feared influenza, terrorism, poison, and a new epidemic.

The building had become part of the disease.

No one yet knew the organism's name.

CHAPTER 3

The Hotel That Became a Habitat

A building designed for people became an amplification system for a microbe no one knew to look for

---

The first patients did not look connected.

One man developed a fever after returning home from Philadelphia. Another became weak and short of breath. A third entered a hospital with pneumonia. Some had attended the same convention. Others had worked nearby or passed through the same part of the city. By the time the pattern became visible, the American Legion's annual convention was over and thousands of people had dispersed across Pennsylvania and beyond.

The convention had opened on July 21, 1976, during the summer of the United States bicentennial. The Bellevue-Stratford Hotel was one of Philadelphia's grand old buildings, a place of chandeliers, ballrooms, meeting rooms, elevators, kitchens, pipes, cooling equipment, and the accumulated modifications of more than seven decades. Legionnaires gathered there for speeches, reunions, meals, and celebrations.

Within days, attendees began falling ill.

The symptoms were alarming: high fever, cough, chest pain, confusion, profound weakness, and severe pneumonia. Initial reports suggested a cluster among veterans who had stayed at or visited the hotel. The death toll rose. Newspaper coverage spread faster than the investigation. The possibilities included swine flu, a toxin, sabotage, heavy metals, contaminated food, and an organism no laboratory had identified.

The Centers for Disease Control sent investigators. David Fraser led the epidemiological work. Teams interviewed patients and families, reconstructed movements, compared people who became sick with those who did not, collected environmental samples, reviewed meals, mapped hotel exposure, and tested blood and tissue against known agents.

The case became a national spectacle because the country had recently prepared for a feared swine-flu pandemic. Every unexplained respiratory death looked like the beginning of something larger.

Weeks passed without an answer.

The investigation narrowed the location. Illness was strongly associated with presence at the Bellevue-Stratford, but no single banquet or food item explained it. People could become sick without staying overnight. The pattern suggested airborne exposure, yet the expected pathogens were absent.

In the laboratory, Joseph McDade had examined tissue from victims using methods intended to find rickettsiae. He did not identify the culprit. Months later, bothered by the unresolved outbreak, he returned to the samples and modified the work. In December, during a period when many colleagues were away for the holidays, he detected rod-shaped bacteria proliferating in guinea-pig tissue. By January 1977, the CDC could announce a new organism.

It would be named *Legionella pneumophila*.

The mystery disease became Legionnaires' disease.

The breakthrough did more than name a pathogen. It revealed a category of habitat that modern society had built without recognizing it as habitat.

The building behind the diagnosis

For much of medical history, the environment of infection was visible. Sewage contaminated water. Mosquitoes bred in standing pools. Food spoiled. Wounds became dirty. Modern buildings pushed many of those hazards behind walls and into engineered systems.

A large hotel has an internal climate. Water enters, warms, cools, circulates, stagnates, evaporates, condenses, and leaves. Air moves through shafts, rooms, equipment, and crowds. Surfaces collect films. Pipes develop scale. Temperature varies by floor, time, use, and distance from heaters. A mechanical system intended to serve people also creates gradients and niches.

Legionella is common in natural freshwater at low levels. The danger arises when it enters a built water system that permits amplification and then produces aerosols capable of reaching susceptible lungs. Warm temperatures, stagnation, sediment, biofilms, and relationships with amoebae can all matter. The bacterium's existence is not the complete hazard. The habitat is.

The exact environmental source in the 1976 hotel was never recovered with the certainty modern outbreak investigators would prefer. Air-conditioning and water-related pathways became central explanations, and later outbreaks established the general mechanism by which building water systems can amplify and aerosolize *Legionella*. The responsible version of the story therefore avoids a false precision: the Bellevue-Stratford was clearly the critical exposure environment; the exact path through that environment remained less certain.

That uncertainty strengthens rather than weakens the lesson.

The system produced a deadly exposure before investigators possessed the vocabulary, sampling strategy, or reference organism needed to trace every step.

The hotel did not merely contain the outbreak.

The hotel participated in it.

The third law: the background is part of the machine

The third law of the Universal Control Layer is:

> ***What a design treats as background can become an active component of the system.***

In a conventional engineering drawing, water is a utility. Air is a flow. A surface is a boundary. A pipe is a conduit. Biology sees additional possibilities: water is a medium; air is transport; a surface is habitat; a pipe is an ecology.

The difference appears in hospitals, food plants, aircraft, data centers, spacecraft, and homes.

A hospital sink can become a reservoir. A decorative fountain can generate aerosols. A warm-water loop can support growth. A cooling tower can distribute droplets across a neighborhood. An air filter can collect nutrients and moisture. A spacecraft surface can be inoculated by the crew and then shaped by cleaning chemistry, humidity, radiation, and material composition.

None of this means every building is dangerous or that sterility is the answer. It means the built environment must be designed and managed as a coupled mechanical-biological system.

Sterility is a temporary state and often an impossible operating goal. Control is the more realistic objective: prevent amplification, interrupt transmission, protect vulnerable people, detect drift, and recover when the system moves outside safe bounds.

The anomaly that almost remained noise

McDade's return to the samples is a story about scientific persistence, but it is also a story about reference dependence.

A diagnostic system identifies what it has been built to recognize. Culture media, incubation conditions, staining methods, antibodies, sequence databases, and software libraries each define a window. A negative result means the target was not detected by that method under those conditions. It does not mean the system contained nothing important.

This is a central problem in metabolomics and microbiome chemistry. High-resolution instruments detect large numbers of features that cannot be assigned confidently to known structures. Some features are artifacts, contaminants, isotopes, adducts, fragments, or redundant views of the same molecule. Others represent real chemistry without a library match.

The temptation is to discard the unknowns so the analysis becomes tractable.

The result is a closed vocabulary: only known signals can enter the explanation, so the unknown can never become causal.

The opposite temptation is equally dangerous: to treat every unknown signal as profound.

The Universal Control Layer requires a middle discipline. Preserve the anomaly, assign a confidence level, test reproducibility, track context, and advance it through an evidence ladder.

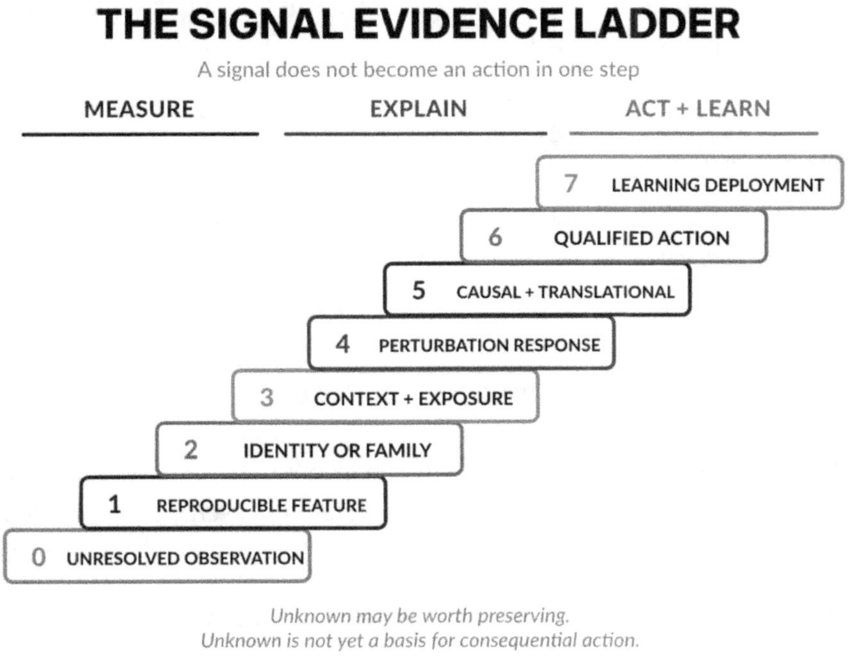

Figure 5. From unknown feature to actionable signal. Detection is followed by quality control, recurrence, contextual association, structural confirmation, exposure, causal testing, intervention response, and independent replication.

A useful ladder might look like this:

Level A - Detected. A signal survives basic analytical quality control.

Level B - Recurrent. It appears reliably across replicates, samples, or time points.

Level C - Context-linked. It changes with a relevant condition, location, perturbation, or outcome.

Level D - Identified. The chemical entity or organism is confirmed with appropriate standards and methods.

Level E - Exposed. The entity reaches the relevant compartment at a plausible concentration and time.

Level F - Causal. Adding, removing, blocking, or rescuing the entity changes the system in a relevant model.

Level G - Actionable. An intervention based on the signal improves an outcome with acceptable risk.

Level H - Portable. The finding replicates across laboratories, populations, devices, and operational settings.

Most signals should never reach the top. That is not wasted effort. A control system earns trust by retiring false alarms as well as discovering real ones.

The cost of finding the answer late

By the time the cause of the 1976 outbreak was identified, the immediate exposure had ended. Investigators could explain the event, but they could not use the answer to prevent the first cases.

This difference separates forensic science from predictive control.

Forensics asks: What happened?

Control asks: What state is developing, and what can be changed before the consequence?

A hospital can test a patient after pneumonia develops. A predictive system would also monitor water temperatures, stagnation, disinfectant levels, biofilm risk, aerosol-generating devices, and environmental *Legionella* before cases appear. The value lies not in replacing clinical diagnosis but in connecting environmental state with human vulnerability.

The same architecture applies elsewhere.

A harmful algal bloom can be identified after toxins appear in drinking water. A control system also watches nutrient loading, temperature, hydrology, community composition, toxin genes, and actual chemistry.

A patient can be diagnosed with sepsis after organ dysfunction. A control system seeks earlier combinations of vital signs, immune signals, metabolites, and clinical context without drowning clinicians in false alarms.

A spacecraft can identify mold after a panel is opened. A control system monitors humidity, condensation risk, volatile products, surface communities, and material changes while preserving manual inspection.

The earlier the intervention, the less force may be required.

That is one of the most important economic and ethical arguments for measurement. Late control tends to be expensive control. It requires more energy, more drug, more disruption, more emergency authority, and more tolerance for collateral harm.

Signal, noise, and the value of the discarded peak

In everyday laboratory language, "noise" can mean several different things.

It can mean random measurement variation. It can mean background signal from the instrument or environment. It can mean a real feature unrelated to the question. It can also mean a real feature whose meaning is unknown.

Those categories should not be collapsed.

When a blood sample enters a high-resolution mass spectrometer, the instrument may report thousands of features. Clinical medicine uses a much smaller set of established analytes. The rest are not all discarded literally, and advanced research programs analyze many of them, but the practical gap remains: a large portion of measured chemistry has no confident structural or functional annotation.

The phrase "off-peak" is useful as an image but imprecise as analytical language. The real opportunity lies in unannotated, low-abundance, transient, or context-dependent features that current workflows cannot interpret reliably.

The Universal Control Layer treats those features as a managed uncertainty reservoir.

They should be retained with metadata:

- sample source and handling;
- time relative to meals, sleep, medication, exercise, and symptoms;
- instrument method and quality controls;
- location and compartment;
- co-varying organisms, proteins, transcripts, and clinical measures;
- response to perturbation;
- confidence in identity;
- evidence for causality.

Over time, an expanding reference engine can convert part of the reservoir into a dictionary.

This is one role for Industrial-Scale Microbiome Mining. ISMM is not a magical translator and not a device that would sit beside every hospital bed. It is an Earth-based discovery and reference capability: recover intact environmental microbial communities at sufficient physical scale, preserve chemical diversity, fractionate and characterize the material, connect features to sources and activities, and create experimentally accessible data for computational and biological interpretation.

The deployed monitor is a different system. It may use mass spectrometry, sequencing, spectroscopy, electrochemical sensors, imaging, environmental probes, or targeted assays. Its interpretation can be informed by the reference knowledge generated through ISMM and other large-scale mapping efforts.

The distinction matters.

The mine builds the map.

The monitor uses the map to interpret the local terrain.

Why the dictionary must include negative meaning

A dictionary of biological state cannot contain only known dangers.

It must also contain normal variation, harmless adaptation, recovery patterns, and conditions in which a signal does not predict harm.

Otherwise every unusual reading becomes a threat.

This is a common failure in early-warning systems. The cost of a missed event is vivid, so thresholds are lowered. Alerts multiply. Operators learn that most alerts do not matter. Attention becomes the scarce resource. The system that was meant to detect danger trains people to ignore it.

The control problem is not maximum sensitivity. It is calibrated decision value.

For each signal, the system should estimate:

- probability that the signal reflects a real state;
- probability that the state will lead to harm without intervention;
- time available before harm;
- likely benefit of acting;
- likely harm of acting unnecessarily;
- value of collecting more information first.

These questions convert detection into decision.

In medicine, the balance is familiar. Screening can save lives when disease has a detectable preclinical phase and early treatment improves outcome. It can also create overdiagnosis, anxiety, invasive follow-up, and treatment of conditions that would never have caused harm. Environmental monitoring can identify an outbreak before patients appear, but poorly calibrated sampling can close facilities unnecessarily or miss local niches.

The Universal Control Layer must therefore include a "do not act yet" state.

More data is not always the excuse of a slow institution. Sometimes it is the correct control action.

Buildings as living extensions of the body

People spend most of their time inside built environments. Those environments shape exposure continuously: air particles, humidity, temperature, light, noise, volatile compounds, water, surfaces, social density, and microbes.

The boundary between habitat and body is porous.

We inhale the building. We drink from it. We touch it. We shed cells into it. We alter its microbial communities, and they alter our exposure. Ventilation changes infection risk. Temperature changes sleep and cognition. Lighting changes circadian timing. Materials release compounds. Water systems select organisms.

A hospital room is therefore not only a container for clinical care. It is part of the patient's physiological environment. A classroom is part of the child's cognitive environment. A spacecraft is part of the astronaut's immune and microbial environment.

This is the larger meaning of the Legionnaires' story. The disease did not respect the categories of architecture, mechanical engineering, microbiology, epidemiology, and medicine. The categories were our administrative convenience. The exposure crossed them without permission.

The Universal Control Layer is an integration discipline for problems that cross categories faster than institutions do.

The habitat-monitoring canvas

A practical habitat control plan can begin with eight boxes.

Purpose. Whom or what is the system protecting?

Critical zones. Where can amplification, exposure, or failure occur?

State variables. Which physical, chemical, biological, and operational variables describe risk?

Sensing plan. Where, how often, and with what confidence will the variables be measured?

Trigger logic. What patterns require observation, sampling, containment, or intervention?

Intervention library. What reversible and irreversible actions are available?

Verification. How will the response and unintended effects be measured?

Fallback. What happens when the sensor, model, network, or actuator fails?

The canvas forces a building owner to move beyond annual compliance. It forces a spacecraft designer to think beyond prelaunch cleanliness. It forces a hospital to connect facility data with clinical risk. It forces the model designer to specify authority and failure modes.

A control layer without a fallback is an automation proposal, not a safety architecture.

The organism no one was looking for

After *Legionella* was identified, investigators could look backward. Stored samples, earlier unexplained outbreaks, and antibodies in human populations revealed that the organism had not appeared suddenly in 1976. The discovery changed the ability to see history.

This is what a new dictionary does.

It does not merely name tomorrow's signal. It reinterprets yesterday's noise.

As the chemical dictionary of the Ghost Genome expands, old biobank samples may acquire new meaning. A feature preserved without identification may later be linked to a microbial pathway, a host response, a dietary precursor, or a disease trajectory. Environmental samples may reveal that a compound appeared before a bloom, crop failure, or outbreak. A patient's longitudinal record may show that the earliest warning was present months before the recognized diagnosis.

The opportunity is profound, but it comes with a requirement: preserve samples, metadata, and uncertainty well enough for future interpretation.

A database optimized only for today's knowns destroys tomorrow's retrospective telescope.

The investigation under pressure

The CDC investigation unfolded in a political environment primed for fear. Earlier in 1976, an influenza outbreak at Fort Dix had raised concern about a possible recurrence of the 1918 pandemic. The federal government launched a national swine-flu vaccination campaign. When American Legion members began dying after the Philadelphia convention, the country wondered whether the feared epidemic had arrived.

Investigators had to work while officials, journalists, families, and the public demanded certainty.

This is the worst environment for an unknown signal. The cost of delay is visible. The evidence is incomplete. Every preliminary hypothesis can alter behavior and trust.

A control layer needs a protocol for uncertainty under public pressure.

It should release what is known, what is not known, what protective action is reasonable, and when the next update will occur. It should separate a working hypothesis from a confirmed cause. It should preserve the ability to change direction without making revision look like incompetence.

The alternative is silence followed by overconfidence, a pattern that destroys trust even when the technical work is sound.

The organism inside another organism

Legionella complicates control because it can live within free-living amoebae in water systems. The amoeba can provide a protected environment in which the bacterium survives and replicates. The relationship helps explain persistence and resistance to environmental stress.

A water sample that counts only free bacteria can miss the protected state. A disinfectant that does not reach the host compartment can appear ineffective. Temperature and biofilm conditions influence the larger ecology.

The unit of control is therefore not one organism. It is a nested habitat: bacterium within amoeba within biofilm within pipe within building within community.

This nesting is common in biology. A virus lives within a cell within a tissue within a person. A tumor clone lives within a microenvironment within an organ. A microbial metabolite exists within a local chemical gradient within a host.

The Universal Control Layer must be able to zoom between levels without pretending one level explains all others.

A building-water state vector

A serious Legionella-control program can illustrate the practical state vector.

It may include:

- water temperature at heaters, returns, distal outlets, and low-use zones;
- disinfectant residual;
- flow, stagnation, and occupancy patterns;
- sediment, scale, and biofilm indicators;
- aerosol-generating devices;

- construction, shutdown, and pressure events;
- targeted organism measurements;
- clinical cases and vulnerable populations;
- corrective-action history;
- confidence and representativeness of sampling.

No one variable controls the risk. A negative sample does not guarantee safety if the sampling plan misses the niche. A positive sample does not have the same meaning in every building. The state must be interpreted against exposure pathways and population vulnerability.

This is why control plans need qualified human judgment even when sensors become continuous.

Preserve the sample, preserve the future

The Legionnaires' investigation succeeded partly because tissue and laboratory material could be revisited. McDade's second look changed the outcome.

Future chemical-state programs should treat sample preservation as infrastructure.

A stored specimen with reliable metadata can be reanalyzed when a new standard, instrument, or hypothesis appears. A destroyed sample leaves only the old interpretation. Biobanks, environmental archives, and mission sample stores become time machines for the control layer.

Preservation has costs and ethical obligations. Consent, privacy, storage, chain of custody, and access must be governed. Not every sample should be kept forever. The decision should reflect future information value.

The unknown signal may become legible years later.

The ability to look again is part of resilience.

From outbreak detection to pre-symptom control

The first three chapters have moved through three failures of apparent control.

Biosphere 2 showed that an inventory is not a state.

Fordlandia showed that simplification can remove invisible protection.

Legionnaires' disease showed that the background environment can become an active machine.

The next step is earlier measurement.

In the early 1960s, a microbiologist named Robert Guthrie began working on a way to identify a metabolic disorder in babies who looked healthy. The disorder could damage the developing brain before the outward signs made the diagnosis obvious. The chemistry was present. The intervention existed. The missing piece was a measurement cheap and simple enough to reach nearly every newborn.

Guthrie turned a few drops of blood on filter paper into a national control system.

PART II

MEASURE FOR PURPOSE

CHAPTER 4

The Baby Who Looked Healthy

A few drops of blood demonstrated that the best moment to intervene can arrive before symptoms

---

The newborn looked normal.

That was the danger.

A baby with phenylketonuria - PKU - can appear healthy at birth. The problem is chemical. A missing or deficient enzyme prevents the body from processing phenylalanine normally. The amino acid and related metabolites accumulate. Without treatment, the developing brain can be irreversibly injured before the family understands what is happening.

By the time the symptoms become unmistakable, the best window for prevention may already be gone.

In the middle of the twentieth century, physicians knew that a restricted diet could protect some children if PKU was recognized early enough. The problem was reach. A diagnostic method that depended on specialized laboratory work after symptoms appeared could not protect babies who had not yet declared themselves sick.

Robert Guthrie wanted a test that could travel.

Guthrie was a physician and microbiologist with an unusual combination of technical persistence and personal urgency. Intellectual disability had touched his family. He became deeply involved in advocacy and in the search for practical ways to identify treatable metabolic disease before damage accumulated.

The test he developed was elegant because it was inexpensive.

A health worker pricked the baby's heel and placed drops of blood on filter paper. After drying, the card could be mailed to a central laboratory. A small disc punched from the blood spot was placed on a culture medium containing a bacterium whose growth was normally inhibited. Elevated phenylalanine overcame the inhibition, creating a visible zone of bacterial growth.

The bacterium became the sensor.

The paper became the transport system.

The mail became the network.

The result became a decision before the child looked ill.

In the early 1960s, Guthrie coordinated a large multistate pilot involving hundreds of thousands of newborns. Massachusetts established the first mandated statewide newborn-screening program in 1963. Other states followed. Over time, the dried-blood-spot architecture expanded far beyond PKU, and tandem mass spectrometry made it possible to screen for many metabolic conditions from one sample.

The Guthrie card was not merely a laboratory test.

It was a population-scale control loop.

The fourth law: the intervention window defines the sensor

The fourth law of the Universal Control Layer is:

> ***Measure early enough for the answer to change the outcome.***

This sounds self-evident. Many systems measure after the useful intervention window because the late state is easier to detect.

A collapsed bridge is easy to identify. Fatigue accumulating in a critical component is harder.

A symptomatic infection is easier to diagnose than the environmental conditions that are preparing the outbreak.

Organ failure is easier to recognize than the distributed metabolic drift that precedes it.

A mature harmful algal bloom is easier to see than the ecological transition that made it likely.

A company crisis is easier to count than the months of decision latency and talent loss that produced it.

The sensor should be designed backward from the last responsible moment to act.

For PKU, the purpose was not to explain intellectual disability after it occurred. It was to prevent avoidable injury. That purpose changed the acceptable timing, price, sample, logistics, and threshold.

A perfect test available only in a research center would not have solved the population problem. A useful test had to be simple enough to reach nearly every birth, stable enough to travel, sensitive enough to identify risk, and connected to confirmatory testing and treatment.

Measurement is not a device. It is a pathway from state to action.

The purpose ladder

The same biological variable can support different purposes.

Phenylalanine can be measured to study metabolism, diagnose PKU, monitor dietary treatment, investigate a variant, or conduct population screening. The desired accuracy, frequency, setting, and consequence differ.

A Universal Control Layer should therefore state the purpose before choosing the measurement.

There are at least five levels of measurement purpose.

Description. What is present, and how does it vary?

Association. What changes with a condition or outcome?

Prediction. What signal improves the estimate of what will happen next?

Diagnosis. What evidence supports a defined clinical or operational state now?

Control. What signal can safely guide an intervention, and how quickly must feedback return?

A measurement can be excellent for description and inadequate for control. An untargeted mass-spectrometry feature may reveal a disease-associated pattern without having enough structural certainty or causal relevance to select a therapy. A wearable can predict fatigue without establishing why it occurred. A microbial abundance can mark a diet without being an intervention target.

The category should be explicit because the proof burden rises as measurement approaches action.

The transition from association to control is where stories can outrun evidence.

Healthy is not one state

Newborn screening also exposes the weakness of the word *normal*.

A population reference range describes a distribution in a selected population. It does not define optimal health for every person. It can include undiagnosed disease. It can vary by age, sex, ancestry, diet, geography, medication, time of day, pregnancy, altitude, and laboratory method. A value inside the range can be dangerous for a particular individual if it represents a rapid departure from that person's baseline.

PKU is powerful as a screening example because the chemistry and intervention are unusually clear. Many future applications of chemical-state monitoring will be more ambiguous.

Suppose a set of low-abundance metabolites changes months before a neurological symptom. The pattern may represent early disease, a medication effect, dietary change, kidney clearance, infection, or harmless adaptation. A system trained on population averages may overreact to individual difference. A system trained only on personal history may miss a slow deterioration that has become the person's new normal.

The Universal Control Layer needs two baselines at once:

- **population context**, which shows how the measurement compares with relevant groups;
- **personal trajectory**, which shows how the individual is changing through time.

The intersection is more informative than either alone.

A personal baseline is also not a single number. It is a range conditioned on sleep, meals, exercise, medication, cycle, season, and health state. The goal is to learn the individual's patterns of variation and recovery.

Measure the trajectory, not only the threshold

Thresholds are administratively convenient. Above this value, act. Below it, do not.

Trajectories reveal more.

A value rising rapidly toward a threshold can be more concerning than a stable value just above it. A patient whose kidney function is declining within the laboratory's normal range may need attention before the result crosses a conventional boundary. A habitat whose oxygen remains acceptable while control effort rises is losing reserve. A microbial community moving steadily toward a hazardous configuration may warrant investigation before the target organism is abundant.

The relevant variables include:

- current level;
- rate of change;
- acceleration;
- variability;
- response to known perturbations;
- time required to recover;
- control effort required to maintain the state.

A trajectory converts the question from *Are we safe?* to *Where are we going?*

This is one reason longitudinal sampling is so important. Cross-sectional studies compare different people at one moment. Longitudinal studies follow the same person or system through time. The latter can reveal transition, resilience, and individual response, but they are harder and more expensive.

The future control layer must make longitudinal measurement easier without pretending that frequent data automatically becomes meaningful.

The "noise" that may contain the warning

A clinical laboratory report typically presents named analytes with validated ranges and interpretation. That discipline protects patients. It also shows how narrow the clinically actionable chemical vocabulary remains compared with the features visible to research instruments.

Untargeted metabolomics can detect thousands of analytical features. A substantial fraction cannot be identified confidently. Some are not distinct molecules. Some are irrelevant. Some are signals of sample handling. Some may be the earliest trace of a process medicine has not learned to name.

The opportunity is not to dump every unknown peak into a medical record.

The opportunity is to build a reference and evidence system that can progressively deconvolute the unknown.

A useful architecture separates four layers.

Raw signal. The instrument's observation, with quality controls and technical uncertainty.

Chemical hypothesis. Candidate formulas, structures, families, and relationships.

Biological context. Source, location, timing, co-varying processes, and exposure.

Decision meaning. Evidence that the signal changes prediction, diagnosis, or response enough to justify action.

Each layer can fail independently. An accurate mass does not prove a structure. A confirmed structure does not prove a source. A source does not prove function. A function in a cell assay does not prove human benefit.

The system should preserve the chain rather than collapsing it into a confidence score no one can inspect.

What ISMM can - and cannot - contribute

Industrial-Scale Microbiome Mining is relevant to this dictionary problem because many microbial natural products exist at concentrations too low for conventional small-volume collection and may be produced only in intact environmental communities. Recovering large amounts of biomass from very large volumes can turn a trace into physical material that can be fractionated, measured repeatedly, structurally investigated, synthesized, and tested.

Scale can therefore be part of the instrument.

But the role must be stated accurately.

ISMM does not directly tell a clinician what an unknown peak in one patient's blood means. Environmental chemistry and human chemistry are not interchangeable. A molecule recovered from an aquatic community may never occur in a person. An environmental natural product can still provide a structure, fragment pattern, biosynthetic clue, target hypothesis, or molecular family that helps interpret related chemistry. It can also supply biologically realized priors for therapeutic design.

The reference engine will require many sources: human cohorts, tissues, longitudinal samples, controlled perturbations, microbial cultures, intact communities, environmental collections, synthetic standards, isotope tracing, clinical outcomes, and negative results.

ISMM is one way to expand the physical chemistry available to that engine.

The deployed diagnostic will be informed by the combined map, not by one company's library alone.

That distinction protects both the science and the commercial claim.

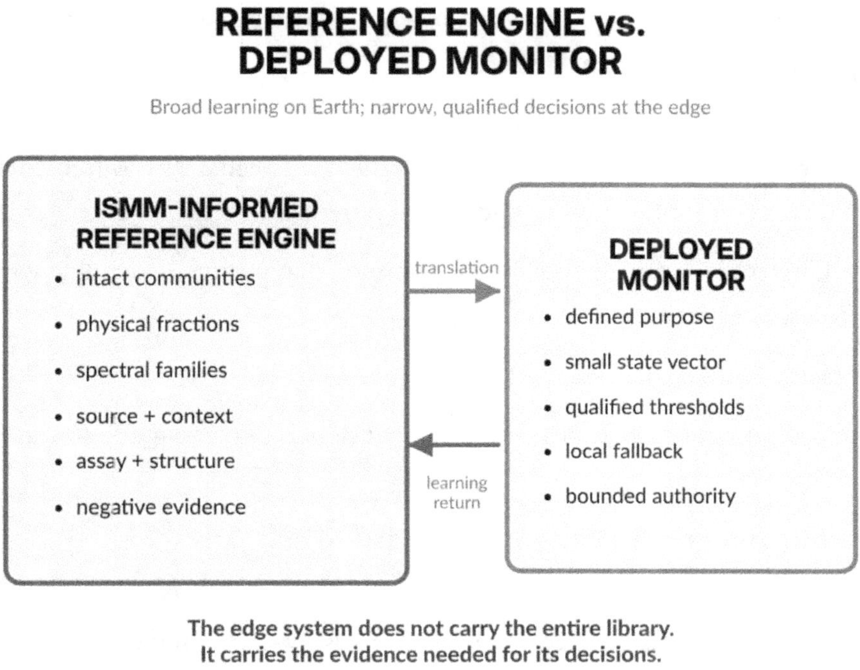

Figure 7. The ISMM reference engine and the deployed monitor. Large-scale Earth discovery expands the chemical dictionary; local sensors in clinics, farms, or habitats compare current state with that broader reference and return new data.

The false comfort of comprehensive measurement

The word *omics* suggests totality: genome, transcriptome, proteome, metabolome. In practice, every omics measurement is partial.

Different extraction methods recover different chemistry. Instruments favor different mass ranges, polarities, charge states, and concentrations. Samples come from accessible compartments. Timing is sparse. Annotation libraries are incomplete. Batch effects can resemble biology. The act of averaging can erase local signals.

"Measure everything" is not an operating plan.

The more useful phrase is **measure enough of the right state, at the right time, for the right decision**.

This is the discipline Guthrie understood. He did not attempt to reconstruct the entire newborn metabolome. He selected a measurement tied to a preventable outcome and built a system around it.

The Universal Control Layer should combine that purpose discipline with exploratory capacity.

A practical system has two tracks:

The operational track measures validated signals that can guide action now.

The discovery track preserves broader data to find signals that may become operational later.

The tracks must communicate without confusing their standards. Discovery can tolerate more uncertainty. Operational control cannot.

This is also where Breakthrough Aggressiveness applies quietly. The discovery track may move rapidly through hypotheses and reversible tests. The operational track must migrate toward stronger validation as the consequence of error rises. The book does not need to repeat the seven-level scale to use its central principle: method follows consequence and maturity.

The ethics of prediction

A system that predicts disease before symptoms creates power and burden.

Who receives the prediction?

What happens when there is no effective intervention?

Can an insurer, employer, school, or government obtain the signal?

How are uncertain findings communicated?

What if the model performs differently across populations?

What if the prediction changes behavior in a way that harms the person?

Newborn screening earned legitimacy because the relationship among test, condition, confirmatory diagnosis, treatment, and benefit became strong enough to justify population action. Even then, programs have debated which conditions belong on panels and how to handle uncertain variants and incidental findings.

A universal chemical-monitoring capability would face these questions at far greater scale.

The governing rule should be purpose limitation: data collected to protect a person or habitat should not automatically become available for unrelated control of that person or community.

The term *control layer* can sound authoritarian. The architecture must therefore include control over the controller.

That means consent where possible, minimum necessary data, transparent models, audit trails, independent oversight, appeal, human judgment, and the right not to convert every prediction into action.

A system capable of early warning is valuable only if the people it protects can trust the institutions operating it.

The value of a cheap, boring sample

Guthrie's test changed medicine partly because the sample was humble.

A few drops of blood dried on paper did not require a refrigerated vial, a specialist at the bedside, or a massive instrument in every hospital. The collection could be standardized. The card could travel. The laboratory could centralize expertise. The result could enter a public-health pathway.

The future Universal Control Layer will also depend on sample architecture.

An ideal signal that requires a risky biopsy every day will not support continuous control. A slightly less informative signal available through breath, saliva, urine, a wearable, wastewater, surface sampling, or a microsample may create greater real-world value.

Sensor design should consider:

- invasiveness;
- sample stability;
- frequency;
- location specificity;
- cost;
- power and maintenance;
- calibration;
- contamination;
- operator skill;
- data transmission;
- privacy;
- action time.

The sophisticated part of a control system is often the logistics that make measurement routine.

Guthrie's personal equation

Guthrie's work was not driven only by abstract public health. His son had an intellectual disability, and a niece was diagnosed with PKU. The family experience helped turn a laboratory problem into a mission of reach.

That personal connection did not prove the test. It clarified the purpose.

Innovation often begins with proximity to a cost others experience statistically. The danger is that urgency can outrun validation. Guthrie's achievement was to convert urgency into a scalable, testable system.

He worked not only on the assay but on the politics of adoption. Screening programs required hospitals to collect samples, laboratories to process them, states to set rules, physicians to confirm diagnoses, and families to receive treatment rapidly. A technical test without the delivery system would have identified babies too late or not at all.

The control loop included legislation, logistics, and follow-up.

This is a useful correction to the idea that technology alone creates impact. The sensor is often the easiest component to celebrate. The unglamorous handoff determines whether the signal changes a life.

False positives, false negatives, and asymmetric harm

Every screening system makes errors.

A false negative misses a child who could have benefited. A false positive subjects a family to fear and follow-up. The acceptable balance depends on prevalence, test performance, treatment benefit, confirmatory testing, and harm of delay.

The logic can be expressed without mathematics.

When missing the state causes irreversible harm and confirmation is available, an initial screen may favor sensitivity. When follow-up is risky, expensive, or harmful, specificity becomes more important. When no intervention exists, the value of prediction becomes ethically complex.

The Universal Control Layer should display the expected consequences of both error types. A single accuracy percentage hides them.

In a spacecraft, a false-positive contamination alert can consume scarce water or shut down a subsystem. A false negative can expose the crew. In an automated drug-delivery device, false action can be immediately dangerous. In environmental policy, a false alarm can destroy livelihoods; a missed signal can create irreversible damage.

The threshold is a moral and operational choice, not merely a statistical one.

The economics of timing

Early measurement has option value.

A signal that arrives before irreversible harm creates choices: repeat, observe more closely, change an input, use a reversible intervention, or prepare resources. The same information delivered after damage has less value even if it is more certain.

This can justify investment in imperfect early-warning systems when their uncertainty is managed.

The economic calculation includes:

- cost of sensing;
- prevalence of the dangerous state;
- value of time gained;
- effectiveness and cost of early intervention;
- harms of false action;
- cost of confirmatory testing;
- downstream cost avoided.

Newborn screening became compelling because the treatment was comparatively simple and they avoided enormous harm.

Many metabolomic predictions will not begin with such a favorable equation. The control layer should prioritize use cases where timing, actionability, and consequence align.

Chemical early warning beyond PKU

Medicine already uses chemistry before symptoms in several ways. Troponin supports recognition of myocardial injury. HbA1c reflects longer-term glucose exposure. Newborn panels detect metabolic disorders. Drug levels guide dosing. Kidney and liver markers reveal organ stress.

The future opportunity is to move from isolated markers toward patterns that represent system transitions.

A pattern might combine host metabolites, microbial products, inflammatory signals, and exposure. It may predict loss of barrier integrity, treatment toxicity, infection risk, or metabolic decompensation. The pattern should not become actionable because an AI can classify it. It becomes actionable when prospective studies show that responding to it improves outcome.

Prediction is not the endpoint.

Improved decision is.

From screening to adaptive monitoring

Newborn screening is a one-time or limited-time population intervention. The Universal Control Layer imagines something more adaptive: repeated measurements interpreted against personal and system baselines.

That future already exists in fragments.

Continuous glucose monitors sample interstitial glucose and reveal trajectories that occasional finger-sticks cannot. Wearables estimate heart rhythm, oxygen saturation, activity, sleep, and temperature. Wastewater surveillance can detect population signals. Environmental sensors can track water chemistry and air quality. Sequencing can identify organisms. Mass spectrometry can profile chemistry.

The missing integration is not simply technical. It is conceptual.

Which signals deserve to be combined?

At what cadence?

Who owns the baseline?

What model is allowed to recommend an intervention?

What is the safe response to disagreement among sensors?

How does the system learn without quietly changing the meaning of the alert?

A control layer needs versioned models and versioned baselines. If the algorithm changes, the historical data must remain interpretable. If the sensor changes, the apparent trend may reflect the device rather than the person. If the population changes, the reference distribution changes.

The chemistry moves. The measurement system also moves.

The fifth question: what would we do differently?

Before adding a sensor, ask:

> ***What action becomes possible because this measurement exists?***

If the answer is unclear, the measurement may belong in discovery rather than operations.

The question does not require an immediate treatment. The action may be to repeat the sample, isolate a subsystem, change observation frequency, collect a different type of data, or alert a qualified human. But there should be a defined path.

Data without decision design creates surveillance, storage cost, anxiety, and the temptation to invent meaning after the fact.

Guthrie's card had a clear path: elevated screen, confirmatory testing, dietary intervention, follow-up.

The next chapter moves from one early-life sample to the most intensely measured year in one astronaut's life. Scott Kelly spent nearly a year aboard the International Space Station while his identical twin, Mark, remained on Earth. Ten research teams followed molecular, physiological, cognitive, and microbial changes across time.

The study did not create a perfect human digital twin.

It revealed why a useful one must be temporal.

CHAPTER 5

One Twin Left Earth

A human baseline became a moving trajectory rather than a single normal range

---

On March 27, 2015, Scott Kelly left Earth.

His identical twin stayed behind.

The image was irresistible: two men with nearly the same inherited genome, one living aboard the International Space Station and the other remaining on the ground. Scott would spend 340 consecutive days in orbit. Mark Kelly, also a former astronaut, would serve as a comparison participant. The arrangement was not a conventional controlled experiment. The brothers had different lives, exposures, histories, and daily environments. The sample size was one pair. Yet the opportunity was unprecedented.

NASA assembled ten research teams. They tracked physiology, cognition, immune response, gene expression, epigenetic signals, metabolites, proteins, microbiome composition, telomeres, and other measures before, during, and after the flight. Blood, urine, saliva, and stool became pieces of a longitudinal portrait. The teams did not ask only what spaceflight changed. They asked which changes reversed after return, which persisted, which moved together, and which might matter for a longer mission.

The results were complicated in the way living systems are complicated.

Some changes were expected. Fluid shifted toward the head in microgravity. Bone and muscle faced unloading. Radiation exposure increased. Sleep and circadian timing were challenged. Diet, elevated carbon dioxide, confinement, workload, and isolation changed the operating environment.

Some findings attracted public attention because they sounded almost mythic. Scott's average telomere length increased during flight, then shortened rapidly after return, leaving an increased number of short telomeres. Gene-expression patterns changed. Immune and metabolic pathways responded. The gut microbiome shifted modestly during the mission and largely moved back toward preflight patterns after landing. Many measured systems demonstrated substantial resilience. Some effects remained altered months later.

The study did not prove what would happen to every astronaut.

It did something more foundational for a control system: it made one person's trajectory visible across multiple layers at once.

The fifth law: baseline is a movie, not a photograph

The fifth law of the Universal Control Layer is:

> ***A useful baseline describes how a system moves, not only where it started.***

A single preflight measurement could not have captured Scott Kelly's ordinary variation, response to launch, adaptation to orbit, response to landing, or recovery on Earth. A population average could not reveal which changes were large for him. Mark Kelly provided valuable comparison, but he was not a stationary copy. His own environment and physiology continued to change.

The relevant object was a time series.

In medicine, the word *baseline* often means the value recorded before treatment or exposure. In control engineering, a baseline must include expected dynamics. A healthy heart does not maintain one rate. It accelerates with exertion and slows during rest. Glucose rises after food and falls. Cortisol follows time of day. Immune markers respond to infection and resolve. A resilient system moves within patterns.

The control question is not whether a value changed.

It is whether the change was appropriate to the perturbation, how much reserve it consumed, whether connected systems changed coherently, and whether the system returned.

A flat line can be as abnormal as a spike.

Perturbation reveals what observation cannot

A system at rest can conceal weakness.

Engineers stress a bridge, cycle a battery, load an engine, and test a pressure vessel because performance under perturbation reveals capacity. Medicine does the same with exercise testing, glucose challenges, stimulation tests, and drug responses.

Spaceflight was an extreme perturbation.

It changed gravity, radiation, schedule, diet, carbon dioxide, physical activity, sleep, isolation, and exposure patterns together. That makes causal attribution difficult. It also makes the study relevant to real systems, where variables rarely change one at a time.

The traditional experimental instinct is to isolate one factor. The systems-control instinct adds a second approach: expose the integrated system to a defined challenge and measure its response across layers.

Both are needed.

Reduction identifies mechanism.

Perturbation reveals coordination.

A future Universal Control Layer should maintain a library of responses to safe, standardized perturbations. How does this person's glucose recover after a meal? How does heart-rate variability respond to sleep loss? How does a microbial community respond to a dietary fiber, antibiotic, travel, or stressor? How does a habitat recover after a humidity excursion or temporary change in crew load?

The response curve may be more informative than the resting measurement.

The state vector becomes personal

Chapter 2 introduced the state vector: the set of variables required to distinguish states that demand different action. The Twins Study showed what happens when the vector becomes personal and temporal.

No single omics layer carried the whole meaning.

A change in gene expression could reflect adaptive response rather than injury. A microbial shift could be driven by diet or environment. A metabolite could connect host and microbial activity. A cognitive change could relate to sleep, stress, carbon dioxide, workload, or other factors. The scientific value came from integration and timing.

A human state vector might include:

- vital signs and rhythm;
- sleep and activity;
- symptoms and cognition;
- blood chemistry;
- immune function;
- metabolic features;
- microbial communities and their outputs;
- medications, diet, and exposures;
- environmental conditions;
- recovery from perturbation;
- confidence and missingness for every signal.

The vector is not a demand to measure everything continuously. It is a design map. It helps identify which variables are needed for the purpose and which can be sampled less frequently or inferred cautiously.

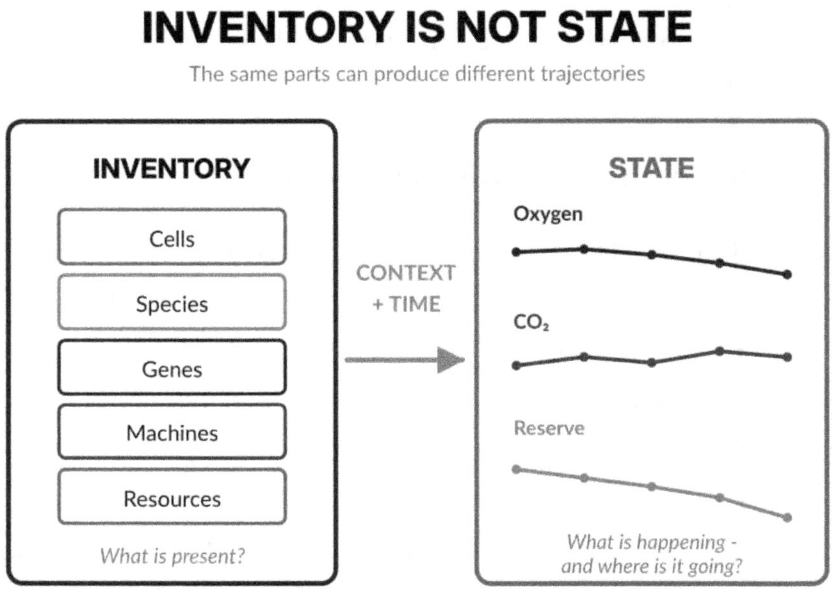

Figure 8. Inventory is not state. A component list describes what exists; a state model adds amount, location, flux, timing, reserve, interaction, and direction of travel.

A good state vector is layered.

Fast variables - heart rhythm, respiration, oxygen, glucose, temperature - can change in seconds or minutes.

Intermediate variables - hormones, inflammatory signals, metabolites, microbial activity, hydration - can change over hours or days.

Slow variables - body composition, bone density, immune repertoire, structural remodeling, long-lived epigenetic patterns - change over weeks, months, or years.

The controller must not sample every layer at the fastest rate. That would waste resources and generate false interpretation. It must understand the time constant of the process.

A spacecraft that waits a month to detect a fast-moving water contamination event is unsafe. A controller that tries to change bone density every hour is absurd.

Time scale is part of biological identity.

The person is not the average astronaut

Population studies are essential for discovering common risks. Mission operations require individual decisions.

One astronaut may be more susceptible to motion sickness, another to vision changes, another to bone loss, another to altered drug metabolism. The same cabin carbon-dioxide level can produce different symptoms. The same exercise program can produce different adaptation. The same food can generate different microbial and metabolic outputs.

Personalization is often marketed as a product category. In control terms, it means something more precise: parameters are updated using the response of the actual system rather than relying entirely on population defaults.

A population model supplies the starting prior.

Personal data update the model.

The controller should not discard the population because the individual is unique. Nor should it erase the individual because the cohort is large. It combines both and shows the uncertainty.

This is Bayesian in spirit even when no formal Bayesian model is used. We begin with what is generally known, observe the person, update, act cautiously, and update again.

The danger lies in overfitting. A model can learn noise in one person's history and treat it as law. Personal baselines should therefore be regularized by physiology, population evidence, and causal understanding.

A twin is not a perfect control.

A personal history is not a perfect model.

The purpose is not perfection. It is a better starting point for the next decision.

Resilience is a measurable phenotype

One of the most reassuring messages from the Twins Study was the body's capacity to adapt and recover. Many changes moved toward preflight levels after Scott returned.

Recovery should not be treated as the absence of a result.

It is a result.

A resilient system detects disturbance, reallocates resources, compensates, repairs, and returns within safe bounds. The speed and completeness of that process can be measured.

For a person, useful resilience metrics might include:

- time to recover sleep after schedule disruption;
- time for inflammatory markers to resolve after infection or exercise;
- metabolic recovery after a meal;
- return of microbial functions after medication;
- cognitive recovery after stress;
- maintenance of function under changing carbon dioxide or workload.

For a habitat:

- time to restore air or water quality;
- ability to isolate a contaminated zone;

- recovery of crop production after disease;
- return of microbial communities after cleaning;
- restoration of material conditions after condensation;
- amount of stored resource consumed during recovery.

The control layer should track not only whether the system returns but how it returns. Overshoot, oscillation, and increasing recovery time can be early warnings.

A patient whose laboratory result repeatedly returns to normal more slowly may be losing reserve before the resting value becomes abnormal.

A habitat requiring larger interventions after similar disturbances is becoming brittle.

Chemical state and electrical function

The language of biological software becomes clearest in the nervous system.

An action potential is commonly described as an electrical signal. That is correct. It is also a chemical and molecular event.

Cell membranes maintain unequal concentrations of sodium, potassium, calcium, chloride, and other ions. Pumps use energy to create and restore those gradients. Ion channels open and close in response to voltage, ligands, mechanical forces, phosphorylation, and other conditions. Neurotransmitters are released, cross synapses, bind receptors, and are removed or metabolized. ATP supports the machinery. Myelin changes signal propagation. Glial cells regulate the environment.

Stop the movement of ions and small molecules, and the electrical signal stops.

The distinction between "electrical" and "chemical" is therefore a difference in description, not a division of life into separate operating systems. Voltage arises from molecular distribution and permeability. Chemistry changes excitability. Electrical state changes chemical release.

This matters to the Universal Control Layer because cognitive performance cannot be monitored or modulated only at the level of behavior. Sleep, nutrition, carbon dioxide, inflammation, medication, microbial metabolites, stress hormones, and ion balance can alter the physical conditions of attention, mood, and decision.

A mission-control system that treats the crew as rational operators outside the biological loop misunderstands the machine.

The human is part of the controlled system and part of the controller at the same time.

From digital twin to adaptive counterpart

The phrase *digital twin* suggests a virtual duplicate that follows a physical system. In aerospace and manufacturing, a digital twin can integrate design, sensor data, maintenance history, and simulation. In medicine, the phrase is used for models that represent aspects of a patient.

The Twins Study warns us not to take the word literally.

A human model will never be a complete duplicate. The body contains unmeasured variables, hidden histories, stochastic events, and interactions at scales no model can resolve in real time. The goal should be an **adaptive counterpart**: a model specific enough to improve decisions, transparent enough to reveal uncertainty, and humble enough to defer when the state lies outside its experience.

The counterpart should answer limited questions:

What trajectory is most plausible?

Which measurements would reduce uncertainty most?

What intervention has the best expected benefit under the current state?

What adverse state should be monitored after intervention?

What would falsify the model?

A model that claims to represent the whole person becomes an authority problem. A model bounded to a purpose becomes a tool.

The AI interpreter

The amount of longitudinal multi-omics and environmental data generated by a mature control layer will exceed unaided human synthesis. Artificial intelligence is necessary for pattern recognition, anomaly detection, integration across time scales, and simulation of alternatives.

It is not sufficient.

AI inherits the boundaries of its data. If unknown chemistry is discarded, the model learns that the known catalog is the world. If the training population is narrow, the model may misinterpret other populations. If treatment decisions in historical data reflect inequity, the model can encode it. If outcome labels arrive late, it can optimize proxies. If the model is rewarded for prediction rather than causal utility, it can identify correlations no intervention should follow.

The Universal Control Layer needs AI with five disciplines.

Provenance. The model should know where each datum came from, how it was measured, and which version of the pipeline produced it.

Uncertainty. It should separate confidence in detection, identity, association, causality, and intervention response.

Counterfactuals. It should compare plausible outcomes under different actions, not merely classify the current state.

Out-of-distribution detection. It should recognize when the current case differs from the data on which it learned.

Auditability. A qualified human should be able to inspect the evidence, assumptions, and changes in the model.

An AI that cannot say "this state is unfamiliar" is unfit for autonomous biological control.

The missingness is data

Longitudinal studies are full of missing samples. Equipment fails. A person is too ill or busy. A specimen is mishandled. A sensor loses contact. A spacecraft has limited freezer space. A laboratory changes methods.

Missingness can be informative.

If measurements are absent during the most stressful periods, the dataset overrepresents stability. If a wearable is removed when symptoms occur, the model may learn that the missing data corresponds to health. If samples cannot be collected during a habitat emergency, the very state requiring control becomes the least observed.

The control layer should track why data is missing and how the missingness changes confidence.

It also needs degraded modes. What decisions remain possible if the high-resolution instrument fails? Can a smaller set of robust measurements keep the system safe? Can the crew perform manual assays? Can environmental compartments be isolated while interpretation is uncertain?

The elegant model is secondary to the resilient operating mode.

What actually changed - and why restraint matters

The Twins Study generated headlines about "space genes" and dramatic telomere changes. The paper itself was more careful.

Spaceflight altered thousands of gene-expression measurements, but gene expression is responsive by design. Many changes moved toward baseline after return. Telomere lengthening during flight was unexpected, yet the postflight increase in short telomeres complicated any simple claim of rejuvenation. Immune responses remained broadly functional, including response to vaccination, while several stress and inflammatory pathways changed. Cognitive measures showed mixed patterns, with some postflight decline that could reflect the demands of return and readaptation.

The microbiome result was modest and transient rather than collapse. Diversity and composition shifted during flight and largely recovered afterward.

This nuance is essential. A control system can be harmed by two opposite errors:

- treating every statistically significant change as dangerous;
- treating recovery of the average as proof that no meaningful cost occurred.

The system must distinguish adaptation, damage, compensation, and recovery.

Multi-omics as a coordination problem

Ten teams produced different data types with different scales and noise. Integration required common time points, metadata, analytical pipelines, and interpretation across disciplines.

This is a preview of the Universal Control Layer's organizational challenge. The biology is integrated before the institution is.

One team owns transcriptomics. Another owns metabolomics. Another owns cognition. Another owns environmental data. Each field has its standards, file formats, confidence rules, and publication incentives. The important signal may sit between them.

Integration cannot mean placing all outputs into one machine-learning model and hoping importance emerges. It requires a causal and temporal scaffold.

Which variables can influence which others?

What is the expected delay?

What is measured directly?

What is a proxy?

Which layer responds first?

Which layer is the outcome?

The scaffold makes the model interpretable and suggests the next experiment.

Resilience versus accommodation

A system can remain functional by adapting to a harmful condition.

An astronaut can perform while sleep-deprived. A patient can maintain blood pressure through compensatory hormones. A habitat can maintain air quality by consuming filter capacity. The visible function persists, but the cost rises.

This is accommodation, not necessarily resilience.

Resilience includes restoration of reserve. Accommodation can become a trap if the stress remains.

The control layer should therefore ask two recovery questions:

1. Did the output return?
2. Did the cost of producing the output return?

A person whose cognition returns only with increased stimulant use, stress hormones, or sleep debt is not fully recovered. A habitat whose water quality returns only through escalating disinfectant has not restored stability.

The second question detects hidden deterioration.

A personal model should know its blind spots

Personalization can create a seductive illusion of precision. A model with thousands of measurements can produce a highly specific recommendation while lacking causal support.

A trustworthy adaptive counterpart should show a **knowledge boundary**:

- variables measured well;
- variables inferred weakly;
- compartments not sampled;
- recent changes in sensor or model;
- populations unlike the current person;
- interventions with limited evidence;
- states never encountered before.

The model should be able to recommend the next measurement rather than the next treatment.

Sometimes the most intelligent action is to ask a better question.

The spaceflight baseline returns to Earth

The Twins Study was conducted for spaceflight, but its logic belongs on Earth.

A person with a chronic disease can be followed through treatment, diet, infection, stress, and recovery. A high-risk worker can be monitored through heat exposure. A patient receiving chemotherapy can be assessed for trajectory rather than waiting for a threshold. A farm can be followed through seasons and interventions. A hospital water system can develop a dynamic baseline. A city can learn its wastewater patterns.

The practical barrier is not only measurement cost. It is the absence of an architecture connecting the data to bounded, ethical decisions.

The Universal Control Layer is that architecture in proposal form.

It does not promise that every change can be understood. It demands that changes be recorded in context, that interpretation be calibrated, and that interventions be linked to verification.

The personal-baseline protocol

A practical protocol for a person or crew member can begin with six steps.

1. **Define the purpose.** Examples: preserve cognitive performance, detect infection early, manage glucose, monitor medication toxicity.
2. **Establish a pre-perturbation window.** Collect enough data to characterize ordinary variation rather than one convenient day.
3. **Record context.** Sleep, food, exercise, medication, symptoms, environment, and major events.
4. **Select sentinel variables.** Use a small operational set for frequent monitoring and a broader discovery set at strategic intervals.
5. **Measure response and recovery.** Focus on trajectories after known challenges.
6. **Set review triggers.** Define when the system increases sampling, alerts a clinician, changes an intervention, or enters a safe fallback.

The protocol is modest compared with the vision of a complete biochemical twin. That is why it can be used.

A control system becomes real through disciplined small loops, not one gigantic model.

The next enclosure

Scott Kelly's body was not the only living system adapting aboard the International Space Station. The station itself had a microbiome. Crew members continuously seeded surfaces and air. Water systems, filters, condensate, cleaning agents, polymers, radiation, elevated carbon dioxide, and microgravity created selection pressures.

The Russian space station Mir had already shown that an orbital habitat could support persistent microbial residents and technical risks. Later studies of the ISS would identify communities that resembled other human-built environments but also reflected the unusual conditions of space.

Sensational accounts would claim that mutated fungi were eating spacecraft alive.

The truth was more useful and more demanding.

Biology was living behind the panels, and the habitat did not yet have a complete way to interpret it.

CHAPTER 6

The Biology Behind the Panels

A spacecraft became an ecosystem whose inhabitants included crew, microbes, condensation, polymers, and time

---

The smell came first.

Cosmonauts aboard Mir sometimes described dampness, stale air, and odors collecting in places that were difficult to inspect. Condensation formed where warm, humid cabin air met cooler structure. Behind panels, around insulation, near seals, and in other protected niches, moisture created small habitats inside the larger habitat.

Mir had been launched in 1986 and grew module by module. It was a technological achievement and an operating improvisation: a long-lived station repaired, expanded, and inhabited far beyond the simplicity of a new vehicle. Human beings continuously brought microbes aboard on skin, in airways, in food, in water, and on cargo. Cleaning removed some. Filters collected some. Dry surfaces limited some. Moisture, nutrients from human shedding, and sheltered spaces supported others.

Over the station's life, investigators isolated hundreds of bacterial and fungal species. A 2004 review reported 234 species identified in the Mir environment. Researchers found organisms capable of persistent colonization and, in laboratory testing, some with potential to degrade polymers or contribute to material corrosion. Fungi such as *Aspergillus* and *Penicillium* appeared frequently. Microbial growth was associated with technical concerns involving surfaces, seals, windows, and equipment.

The dramatic retelling says fungi mutated in space and began eating titanium.

The evidence supports a more careful and more important conclusion.

Spacecraft are not sterile cans. They are human-built ecosystems. Microbes introduced from Earth can persist, adapt, form communities, interact with materials, and create both medical and technical risks. Moisture and material design can matter as much as species names. The hazard emerges from the coupled state.

That state may remain hidden behind a panel until the smell, stain, short circuit, corrosion, or illness makes it visible.

The sixth law: the habitat is part of the organism

The sixth law of the Universal Control Layer is:

> ***In a closed mission, the habitat and its inhabitants form one coupled physiological system.***

This does not mean a spacecraft is literally alive. It means the crew's health cannot be separated cleanly from air, water, surfaces, food, waste, light, temperature, microbes, and materials.

The crew changes the habitat.

The habitat changes the crew.

Every breath alters carbon dioxide and humidity. Every shower, meal, exercise session, skin cell, cleaning event, and waste process changes the environmental state. The station's materials absorb, release, and react. Filters select. Disinfectants select. Radiation damages cells and materials. Microgravity changes fluid behavior and human physiology. Elevated carbon dioxide can influence cognition and symptoms. Dietary constraints alter both human and microbial metabolism.

The usual engineering diagram places the human in a box labeled "crew" and the vehicle in a box labeled "life support." The control problem runs through both boxes.

A water-system biofilm can affect health and equipment. A medication can alter a crew member's microbiome, which changes shedding into the habitat. A cleaning chemical can suppress one community and favor another. A dietary change can alter waste composition and bioprocessing. A material selected for weight and strength can become a microbial substrate under conditions no ground test reproduced.

The boundaries remain useful for responsibility. They are not reliable boundaries for causality.

From Mir to the ISS: less drama, better data

The International Space Station inherited the lesson and added better monitoring.

Studies of the ISS have found microbial communities shaped strongly by human occupancy, resembling those in other confined built environments while fluctuating with crew, location, cleaning, and time. Surface sampling, air monitoring, culture, sequencing, and genomic analysis have expanded the view beyond organisms willing to grow in one laboratory condition.

In 2015, strains of *Enterobacter* were isolated from surfaces associated with the station's toilet and exercise platform. Later genomic work identified strains related to *Enterobacter bugandensis*, an opportunistic pathogen known from clinical settings. A 2024 study compared thirteen ISS-derived strains with Earth counterparts and reported genetic and functional differences associated with persistence, stress response, and antimicrobial resistance.

The finding was important. It did not prove that the organisms had infected an astronaut. It did not establish that space had created a superbug destined to escape. It showed that a closed, human-built environment can support persistent strains with traits worthy of surveillance.

This is the correct tone for control engineering: neither dismissal nor apocalypse.

A signal becomes a reason to improve the loop.

When material becomes habitat

A spacecraft designer ordinarily asks whether a material can carry load, survive vibration, resist flame, tolerate radiation, off-gas within limits, and remain stable across temperature cycles. A biological control layer adds another set of questions.

Does the surface hold a microscopic film of water?

Does it trap skin oil, food residue, or cleaning compounds?

Does radiation or aging expose new chemical groups that make attachment easier?

Can an organism use an additive, coating, plasticizer, sealant, or contaminant as carbon?

Does one material create an electrochemical relationship with another in the presence of microbial metabolism?

The answers are rarely contained in a conventional material specification. A polymer that is durable in dry terrestrial testing can behave differently under persistent humidity, crew-derived nutrients, microgravity-driven fluid distribution, repeated disinfection, and years of inaccessible use. A metal does not have to be eaten as food for biology to accelerate damage. Microbes can create local acidity, alter oxygen, concentrate ions, build matrices, and change electrochemistry at the surface. The organism, its products, the moisture layer, and the material form a reaction zone.

This is the kind of problem that disappears when every discipline takes home only its own sample.

The materials team sees pitting.

The microbiology team sees a mixed community.

The environmental team sees condensation.

The maintenance record shows a hidden compartment that was rarely opened.

The control layer begins when the four observations are treated as one event.

A useful failure investigation would preserve the location, material batch, temperature history, humidity duration, cleaning history, electrical conditions, visible growth, chemical products, and community composition. It would ask not simply whether a species was present, but whether the combined state could plausibly produce the observed damage. It would then recreate the state, not merely culture the organism alone.

That sounds expensive. Replacing a failed subsystem in deep space is more expensive.

The sampling paradox

The act of looking can erase the condition we need to understand.

A panel is opened. Air moves. Moisture evaporates. Oxygen enters. Temperature changes. A swab collects material from the most accessible surface rather than the active boundary. The sample is placed in a container, transported, frozen, thawed, extracted, and analyzed under conditions unlike the niche from which it came. Each step may be necessary. Each step also edits the evidence.

This is the same paradox that complicates human diagnostics and environmental chemistry. A blood draw captures a moving system after the needle, tube, anticoagulant, delay, temperature, and analytical method have already begun to select what remains visible. A stool sample favors chemistry that survived to the end of the gut. A bulk-water sample averages away a biofilm boundary. A surface swab may collect cells but miss volatile or rapidly degraded signals.

The engineering answer is not to reject sampling. It is to design a **sampling chain of custody for state**.

That chain records:

- the exact place and time;
- the environmental variables immediately before collection;
- whether the system was disturbed;
- which fractions were preserved for culture, sequencing, chemistry, imaging, and materials analysis;
- which compounds or organisms may be lost under the preservation method;
- what the instrument can and cannot detect;
- how the signal compares with a personal or habitat baseline.

A mature monitor also knows when a destructive sample is unnecessary. Continuous humidity, volatile-compound, optical, acoustic, electrical, or electrochemical sensing may reveal a change before anyone opens the compartment. The first alert can trigger a carefully planned confirmatory collection rather than a hurried swab that destroys the scene.

The forensic principle is simple: preserve the system's story before dismantling the system.

Planetary protection runs in both directions

The microbial control problem does not stop at the spacecraft wall.

Space programs have long worried about **forward contamination**: carrying Earth organisms to another world and confusing the search for indigenous life or altering an environment before we understand it. Sample-return missions add the reverse concern: bringing material back to Earth without adequate containment and analysis. Crewed exploration complicates both because human beings cannot travel without their microbial companions.

The Universal Control Layer does not resolve planetary-protection policy. It clarifies the measurement problem beneath it.

To know whether a detected organism or molecule belongs to the destination, the vehicle, the crew, or a transformation among them, we need a much better departure inventory than a simple list of cultured contaminants. We need strain-level records where appropriate, chemical fingerprints, material blanks, temporal baselines, and the ability to distinguish a terrestrial molecule from a locally transformed product. We also need governance that decides what evidence permits an action with planetary consequence.

A future Mars habitat may deliberately use microbes to recycle waste, make food, produce materials, or manufacture medicines. The same mission may be trying to protect an external sample from terrestrial contamination. The biological system will therefore contain zones with opposite purposes: cultivate here, exclude there, monitor the interface between them.

That is not a sterilization problem. It is a boundary-control problem.

The sensationalism penalty

Emerging science often travels through a media filter that rewards the most vivid version.

“Microbes detected on ISS surfaces” becomes “space bacteria are mutating into superbugs.”

"Fungi show polymer-degradation potential" becomes "mold is eating the spacecraft."

The amplification can attract attention and funding. It can also damage the control system.

If early signals are exaggerated, operators learn to distrust warnings. If uncertainty is hidden, the later correction looks like retreat. If a potential hazard is described as a demonstrated catastrophe, resources can be misallocated away from more probable risks.

The Universal Control Layer needs a communication protocol that preserves consequence without sacrificing calibration.

Every alert should answer:

- What was directly observed?
- What is inferred?
- What remains possible but unproved?
- What action is justified now?
- What additional evidence would change the action?

This protocol is especially important in autonomous systems. A crew cannot afford to treat every unknown organism as an emergency, nor can it afford to dismiss persistent changes because earlier headlines were overstated.

The reference engine and the local monitor

A habitat cannot carry every instrument, specialist, and library available on Earth. It needs a layered architecture.

At the base is the **reference engine**: large, distributed, continually updated knowledge built from Earth laboratories, environmental collections, clinical cohorts, material tests, model systems, and missions. Industrial-Scale Microbiome Mining can contribute physical chemistry and source context to this engine. Sequencing initiatives, metabolomics repositories, microbial culture collections, materials-testing programs, and medical data contribute other layers.

Above it is the **mission knowledge package**: the subset relevant to the habitat's materials, crew, food systems, water architecture, expected exposures, and intervention capabilities.

At the edge is the **deployed monitor**: rugged sensors and assays capable of operating with limited mass, power, consumables, bandwidth, and expertise.

Above the monitor is the **local interpreter**: software that compares observations with baseline, recognizes anomalies, requests confirmatory tests, and recommends bounded actions.

Above the interpreter is **human authority and safe fallback**.

The loop returns mission data to the reference engine when communication permits, improving the map for future missions.

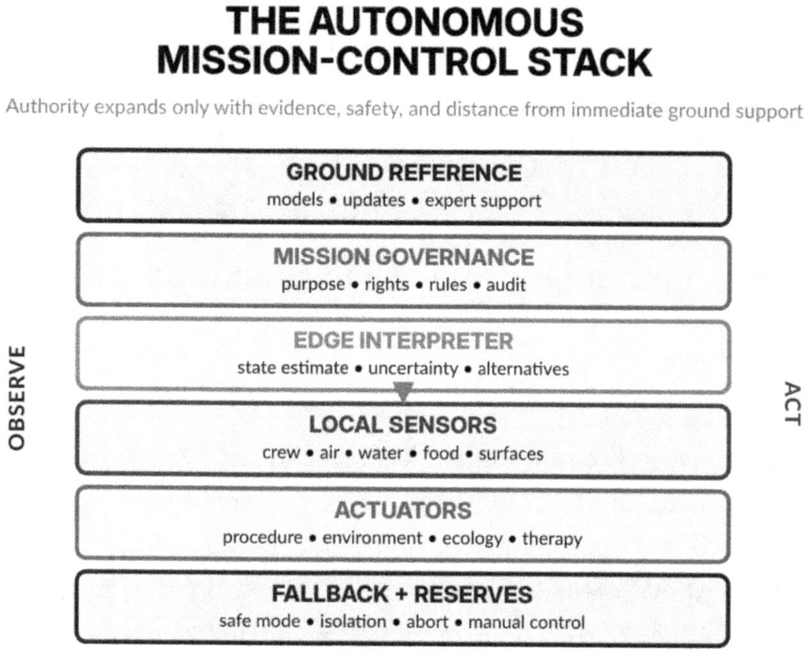

Figure 9. The autonomous mission-control stack. Earth reference knowledge informs a mission package; local sensors and interpreters operate within bounded authority; human override and safe fallback remain available; mission data updates the Earth engine.

This architecture avoids the mistaken statement that ISMM itself must be deployed as a continuous monitor. The mining platform creates reference material and knowledge. The spacecraft needs the monitoring capability that such knowledge helps inform.

What the sensor should smell

Human beings possess crude but useful anomaly detectors. Smell can reveal microbial growth, overheating electronics, leaks, smoke, and chemical release. On Mir, odor and visible growth sometimes prompted inspection of hidden areas.

An engineered system needs more specific equivalents.

For microbial and material risk, useful signals may include:

- humidity and condensation duration;
- temperature at hidden surfaces;
- water flow and stagnation;
- disinfectant residual;
- volatile organic compounds;
- particle counts;
- corrosion products;
- changes in electrical resistance;
- microbial abundance and community composition;
- targeted pathogen assays;
- metabolic fingerprints;
- material mass, optical, or mechanical change.

The best early-warning signal may not be the organism itself. It may be the environmental condition that gives the organism opportunity.

A humidity sensor behind a panel can be more actionable than a species list taken after damage. A volatile metabolite pattern can indicate growth before visible biomass. A change in water chemistry can trigger flushing before a biofilm matures.

The Universal Control Layer should therefore combine **organism sensing**, **activity sensing**, and **habitat sensing**.

Identity tells us who may be present.

Activity tells us what chemistry is occurring.

Habitat tells us whether amplification is likely.

The false promise of total sterility

Spacecraft are assembled with strict contamination controls, especially when planetary protection matters. Medical systems rely on sterilization. Food and pharmaceutical manufacturing require defined microbial limits. These practices save lives and protect science.

A permanently sterile crewed habitat is not realistic.

People are microbial ecosystems. Supplies arrive from Earth. Organisms occupy inaccessible niches. Some microbes are harmless or beneficial. Aggressive sterilization can damage materials, consume resources, select resistant survivors, and eliminate competitors that constrain opportunists.

The operational goal should be **managed ecology within safety bounds**.

That includes:

- keeping high-risk compartments and processes appropriately sterile;
- preventing moisture and nutrient accumulation;
- monitoring high-consequence organisms and activities;
- maintaining cleaning methods with different failure modes;

- preserving safe microbial functions where relevant;
- isolating and remediating local growth;
- avoiding unnecessary antimicrobial pressure;
- verifying recovery.

The phrase managed ecology can sound permissive. It is more demanding than indiscriminate cleaning because it requires understanding the system after the intervention.

A surface that tests clean immediately after disinfection may be entering a less stable state. The important question is what recolonizes, from where, and with what function.

The control problem of biofilms

Biofilms are a useful model for why identity alone is insufficient.

In a biofilm, microorganisms attach to a surface and build a matrix containing polymers, proteins, extracellular DNA, minerals, and other materials. The matrix changes transport, chemical gradients, stress response, and susceptibility to treatment. Cells in the biofilm can behave differently from the same organism growing freely in a flask.

A water line can therefore contain a protected community even when bulk-water samples appear acceptable. Disinfectant concentration at the inlet may not reflect exposure inside the matrix. A culture-negative sample may not prove absence.

The control layer needs spatial reasoning.

Where are the low-flow zones?

Where does temperature enter the growth range?

Where do nutrients accumulate?

Where can aerosols form?

Which surfaces are difficult to inspect or replace?

Which material changes indicate matrix development?

The state is not the average of the system. It may be the worst local niche connected to a vulnerable pathway.

The data problem at the edge

A deep-space habitat will generate more data than it can send continuously to Earth and face communication delays that make immediate ground control impossible. Mars communication delay varies with planetary geometry and can reach many minutes one way. A medical or environmental decision may need to be made before an Earth team can respond.

The deployed interpreter therefore requires autonomy.

Autonomy creates a trust problem.

The model must distinguish routine adaptation from dangerous drift. It must function when sensors disagree. It must resist corrupted data and cyberattack. It must preserve human attention for the few alerts that matter. It must explain why it recommends an action. It must know its own operating envelope.

A useful design divides autonomy into tiers.

Tier 0 - Observe. Collect and display data without recommendation.

Tier 1 - Advise. Recommend additional sampling or inspection.

Tier 2 - Contain. Take reversible protective actions, such as isolating a loop, increasing ventilation, or changing sampling frequency.

Tier 3 - Modulate. Adjust a biological or chemical process within validated bounds.

Tier 4 - Emergency action. Take high-consequence action under predefined conditions when delay is more dangerous than error.

Authority should rise only as evidence, validation, and fail-safe capability rise.

The system should also be able to move down a tier when sensors degrade or the state becomes unfamiliar.

The safe fallback is part of the design

A control system will fail.

The question is how.

Sensors drift. Reagents expire. pumps clog. software updates introduce errors. Models encounter unfamiliar states. Communications fail. Crew members disagree. A local intervention has an unexpected side effect.

The Universal Control Layer requires a fallback plan for each critical loop.

For habitat microbiology, fallback might include:

- isolate the affected water or air subsystem;
- switch to stored reserves;
- increase manual sampling;
- use a broader but well-understood remediation;
- reduce crew exposure;
- send compressed data to Earth;
- preserve samples for later analysis;
- restore operation gradually with verification.

For a patient, fallback might include switching from automated dosing to clinician-controlled mode, using a conservative therapy, repeating a gold-standard test, or stopping an experimental intervention.

Fallback is not an admission that autonomy is weak. It is evidence that the system understands consequence.

The habitat as a learning partner

Every mission creates a new environment. It should not be treated only as a deployment site for Earth knowledge. It is a scientific partner.

The habitat can reveal:

- which Earth microbial functions persist in microgravity;
- how materials and communities interact over years;
- how closed diets alter chemistry;
- which signals predict loss of resilience;
- how human and environmental microbiomes exchange;
- which interventions preserve safety without destabilizing the ecosystem;
- how local production changes exposure and waste.

The mission data should return with full context. A sequence without environmental metadata is incomplete. A metabolite without location and time is incomplete. An intervention without pre- and post-state is incomplete.

The system learns only when the loop is closed scientifically as well as operationally.

From sensing to action

The first half of the book has focused on seeing.

Biosphere 2 showed why state matters.

Fordlandia showed why context matters.

Legionnaires' disease showed why the built environment matters.

PKU screening showed why timing matters.

The Twins Study showed why trajectory matters.

Mir and the ISS show why sensing must be coupled to a habitat model and safe autonomy.

The next question is intervention.

Once a system detects drift, where should it act?

The answer is not always at the molecular target. Sometimes the strongest intervention is environmental. Sometimes it is behavioral. Sometimes it is ecological. Sometimes it is a drug. Sometimes it is a coordinated set of actions across levels.

At the beginning of the twentieth century, engineers trying to cut a canal across Panama had machines capable of moving earth. The project was being defeated by living vectors and the diseases they carried.

The canal would not be built by excavation alone.

It first required a control action upstream of the shovel.

PART III

MODULATE FOR PURPOSE

CHAPTER 7

The Canal Medicine Built

Mechanical scale became possible only after the disease system was changed upstream

---

The machines were not the first thing to fail in Panama.

The people did.

When the French began their great attempt to cut a canal across the Isthmus of Panama in the 1880s, they brought ambition, capital, engineering experience, and the prestige of Ferdinand de Lesseps, whose leadership had helped create the Suez Canal. Panama was not Suez. Tropical rain filled excavations and triggered landslides. Rivers surged. Equipment broke. Management deteriorated. Yellow fever and malaria moved through the workforce.

Thousands died.

The mosquito did not appear in the engineering plan because the disease model was wrong. Yellow fever was commonly attributed to filth, vapors, or direct contagion. Malaria carried its own older theories. The insects were treated as nuisances, not as components of a transmission system.

A Cuban physician, Carlos Finlay, had argued beginning in the nineteenth century that a mosquito carried yellow fever from person to person. His hypothesis was ridiculed and ignored for years. At the turn of the century, Walter Reed's U.S. Army commission, building on Finlay's insight and the work of colleagues, produced experimental evidence supporting mosquito transmission. Ronald Ross had connected mosquitoes with malaria transmission through parasites in another vector species.

The map changed.

William Crawford Gorgas had applied mosquito-control methods in Havana. When the United States took over the canal effort in 1904, he became chief sanitary officer in the Canal Zone. He asked for personnel, money, screens, fumigation, drainage, oil, water systems, waste removal, housing improvements, and authority.

To many engineers and administrators, this looked like delay.

They had come to dig.

Gorgas wanted to change the habitat before the shovels could win.

His teams screened buildings, fumigated rooms, covered water containers, drained breeding sites, applied oil to standing water, cut vegetation, improved water and sanitation systems, isolated yellow-fever patients from mosquitoes, distributed quinine, and organized inspection. The campaign was relentless and intrusive. By late 1905, the last yellow-fever death associated with the canal workforce had occurred. Malaria proved harder because its mosquito vectors behaved differently, but illness and mortality declined substantially as control improved.

The canal opened in 1914.

The standard heroic version says Gorgas conquered disease and made the engineering possible.

The complete version must say more. The campaign stood on Finlay's long-dismissed insight and on experiments involving human subjects whose ethics remain contested. U.S. rule in the Canal Zone was colonial and racially stratified. Health, housing, pay, and protection were distributed unequally. Sanitation was used not only to protect life but also to discipline populations and support an imperial project.

A control system can produce a desired technical outcome while violating human purpose.

That is why the Universal Control Layer begins with purpose and governance, not merely effectiveness.

The insight before the infrastructure

The Panama story is sometimes told as though one decisive American administrator arrived with the correct answer. The actual chain is more revealing.

Carlos Juan Finlay had spent years studying yellow fever in Cuba. In 1881 he publicly proposed that a mosquito served as the intermediary between a sick person and the next victim. He even identified *Culex*—later classified as *Aedes aegypti*—as the likely vector. The claim did not fit the dominant disease models, and Finlay could not yet produce evidence strong enough to reorganize public health around it. For almost two decades, his idea occupied the uncomfortable space between informed hypothesis and operational truth.

After the Spanish-American War, the U.S. Army Yellow Fever Commission led by Walter Reed tested mosquito transmission in Cuba. The experiments helped establish that the disease was not spread through ordinary contact or contaminated bedding in the way many believed. They also involved human volunteers, financial inducements, military hierarchy, and ethical practices that would be evaluated very differently today. The scientific advance and the moral complexity belong in the same account.

Gorgas then translated the new causal map into operations. He did not discover the mosquito hypothesis, and the campaign was not one dramatic spray. It was an organized system of inspection, screens, source reduction, patient isolation from mosquitoes, fumigation, drainage, water management, quinine, housing, and enforcement. The intervention worked because it attacked several links in the transmission cycle and because someone had the authority to sustain the work long enough for the results to become visible.

This sequence—insight, causal proof, operational design, sustained execution—is easy to compress after success. In real time, each step required a different proof burden and a different kind of leadership.

Finlay needed permission to be taken seriously before the entire mechanism was settled.

The Reed Commission needed disciplined experimentation.

Gorgas needed resources and authority to act on a model that engineering leadership initially treated as peripheral.

The canal needed all three.

The lesson for a Universal Control Layer is that the actuator often exists before the institution is ready to use it. Screens, drainage, and mosquito control were not technologically exotic. What was missing was a validated causal model and a management architecture capable of moving the intervention upstream.

Choosing the real control point

A control point is not simply the place where an effect can be produced. It is the place where an intervention can change the trajectory with acceptable cost, evidence, authority, and collateral damage.

Yellow fever offered several possible points.

Treat the fever after infection.

Protect individual workers with barriers.

Reduce contact between patients and mosquitoes.

Destroy adult mosquitoes.

Remove breeding water.

Change housing and sanitation.

Each acted at a different point in the cycle. None was sufficient in every circumstance. The campaign succeeded through a portfolio whose components reinforced one another.

Modern medicine often begins at the molecular target because the target is legible to the discovery system. But the highest-leverage control point may be farther upstream or downstream.

A receptor can be blocked, but the exposure causing repeated activation may be removable.

An inflammatory signal can be suppressed, but a barrier defect or persistent infection may be sustaining it.

A pathogen can be killed, but a device, water system, or behavior may keep reintroducing it.

A metabolic pathway can be altered, but sleep, diet, medication, or environment may be the controllable driver.

Conversely, upstream intervention can be impossible or too slow. A patient in shock needs immediate physiological control while the source is being found. A crew facing toxic air cannot wait for a complete ecological model. Control architecture therefore separates **stabilization**, **source control**, and **restoration**.

Stabilization keeps the system within survivable bounds.

Source control interrupts the process producing the dangerous state.

Restoration rebuilds function and resilience after the threat is contained.

The mistake is to confuse the first successful action with the complete solution.

The counterfactual test

Before acting, a team should be able to describe the counterfactual.

What happens if we do nothing?

What happens if we act only at the symptom?

What happens if we remove the suspected source?

What happens to other functions that depend on the same pathway?

How quickly should the state change if the model is correct?

In Panama, a decline in yellow-fever transmission after vector control strengthened the causal and operational model. In a patient, a molecular marker may change without the person improving. In a habitat, microbial abundance may fall while corrosive chemistry persists. The control target and the outcome target are not always the same.

A disciplined intervention therefore specifies a chain:

action → immediate mechanism → intermediate state → human or system benefit.

Every arrow can fail.

The actuator may not reach the intended site. The mechanism may be wrong. The intermediate state may not matter. The benefit may be offset by another harm. A control layer that monitors only the first arrow can declare success while the system continues toward failure.

Control debt

Organizations accumulate technical debt when expedient solutions make future change harder. Living systems accumulate a related burden that can be called **control debt**.

A broad antimicrobial suppresses a dangerous organism but selects resistance and alters the surrounding community.

A pesticide protects one harvest but reduces ecological redundancy.

A sedating medication makes a short mission task tolerable but impairs judgment later.

A highly restrictive diet simplifies logistics but narrows microbial and nutritional resilience.

An emergency oxygen injection keeps people alive but masks a carbon-flow model that remains incomplete.

These actions can be correct. Debt is not the same as error. Debt means the intervention creates a future obligation: monitor, repay, compensate, redesign, or accept a known loss of resilience.

The intervention brief should therefore include not only expected benefit and immediate risk, but the debt it may create. Who owns the follow-up? What signal reveals that the debt is growing? What restoration action is available?

A system that acts quickly and never pays its control debt becomes progressively less controllable.

The seventh law: intervene at the highest leverage point you can govern safely

The seventh law is:

> ***Act upstream when the upstream cause is known, actionable, and ethically governable.***

The French effort attacked the visible work: move earth faster, replace workers, repair equipment. Gorgas attacked the transmission conditions that were removing the workforce.

Upstream action can be more efficient because it prevents many downstream consequences at once.

Drain a breeding site and prevent multiple bites.

Remove a contaminated source and prevent multiple infections.

Change a habitat condition and suppress amplification before treatment is needed.

Correct a dietary deficiency and prevent a cascade of symptoms.

Repair a process that produces defects rather than inspecting defects at the end.

But "upstream" can become a slogan that justifies excessive authority. Population-level intervention can be coercive, inequitable, and based on an incomplete model. The fact that a leverage point exists does not establish who has the right to use it.

The Universal Control Layer therefore applies three tests.

Causal confidence. Is the upstream factor truly contributing to the outcome, or is it merely associated?

Controllability. Can the factor be changed with adequate reliability and specificity?

Legitimacy. Does the actor have ethical and legal authority, and are affected people protected and represented?

A technically elegant intervention that fails the third test is not a successful control system.

The intervention ladder

Biological systems can be modulated at several levels. The best starting point is often the least invasive level capable of changing the trajectory.

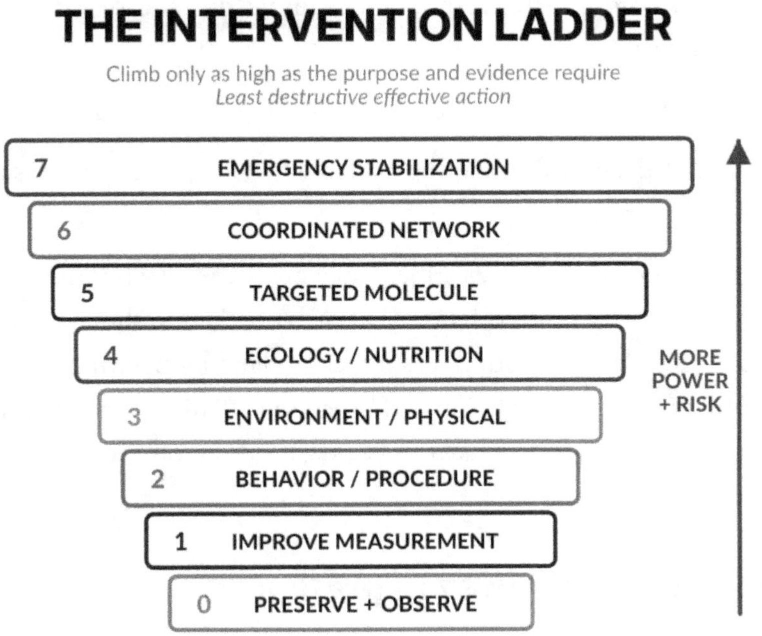

Figure 10. The intervention ladder. Begin with purpose and consider progressively more direct actions: information and behavior, environment, inputs, community or vector, pathway, molecule, device, and structural redesign. Escalate only when needed and verify at every step.

A practical ladder includes:

Information and behavior. Change timing, practice, exposure, adherence, workflow, or decision.

Environment. Change temperature, humidity, flow, ventilation, light, surface, sanitation, or spatial structure.

Input. Change food, nutrient, precursor, water, oxygen, substrate, or material entering the system.

Community or vector. Change microbial ecology, insect transmission, competition, colonization, or reservoir.

Pathway. Alter a biological process through a targeted intervention.

Molecule or combination. Add, block, mimic, degrade, or sequester specific chemistry.

Device or automated loop. Couple sensing to controlled delivery.

Structural redesign. Rebuild the system when local interventions cannot produce a safe state.

The ladder is not a rigid order. Emergency medicine may require immediate molecular or mechanical action. A patient in shock does not wait for a lifestyle intervention. A spacecraft water loop with a dangerous contaminant may need isolation and broad remediation. The discipline is to ask whether a less destructive upstream action can prevent repeated downstream force.

The target is not always a protein

Drug discovery often uses the word *target* to mean a protein or other molecular entity that a compound can engage. That definition is useful and will remain central to medicine.

The control target can be broader.

For yellow fever in Panama, the control target was the transmission loop among infected human, mosquito, susceptible human, and breeding habitat. The molecular virus mattered, but the available leverage point was ecological.

For *Legionella*, the target can be water-system conditions as well as the bacterium.

For obesity, meaningful control may involve appetite pathways, gastric signaling, food environment, sleep, activity, muscle, liver, adipose tissue, and microbial metabolism. A powerful drug target can still be part of a larger control architecture.

For antibiotic resistance, the target can include prescribing patterns, diagnostics, infection control, agricultural use, economic incentives, and molecular mechanisms.

A protein target tells us where a molecule can bind.

A control target tells us where an intervention can change the outcome.

The two can coincide. They should not be confused.

Modulation is not domination

The word *modulate* is deliberately weaker than *command*.

Living systems compensate. A strong intervention can activate counter-regulation, select escape variants, or damage a function connected to the same pathway. Complete inhibition can be less effective than partial adjustment. Local action can be safer than systemic exposure. Timing can matter more than peak potency.

Modulation asks:

- How much change is needed?
- Where is it needed?
- For how long?
- In which direction?
- What other nodes will respond?
- What happens when the intervention stops?

This is the logic behind feedback dosing, combination therapy, pulsed treatment, local delivery, and adaptive management.

It is also the logic of environmental control. Gorgas did not eliminate all mosquitoes from Panama. He interrupted transmission sufficiently for the project and population context. The intervention was intensive, but the biological objective was bounded.

The strongest control often comes from changing a probability rather than eliminating an entity.

The forgotten control: stopping the actuator

Engineers devote enormous attention to how an actuator starts and less attention to how it stops safely.

A biological intervention can continue after the decision has changed. A drug remains in the body. A gene-editing event may be irreversible. An introduced organism can reproduce. A pesticide persists. An automated infusion can overshoot because of sensor delay. An ecological intervention can spread beyond its intended zone.

The Universal Control Layer requires a stop design.

For every intervention:

- What is its duration?
- Can it be reversed?
- Can it be localized?
- What is the maximum exposure?
- What condition forces cessation?
- What residual effects remain?
- What rescue exists if the response is wrong?

Reversibility is a form of information value. A reversible intervention allows the system to learn while limiting downside. An irreversible intervention requires stronger evidence before action.

This is where the BAS idea of matching aggressiveness to consequence becomes operational. The same exploratory freedom appropriate for a simulation is not appropriate for a self-propagating biological release.

Measure for purpose, modulate for purpose

The phrase at the center of this book is:

> ***Measure for purpose so that we can modulate for purpose.***

Measurement without purpose creates data accumulation.

Modulation without measurement creates blind force.

Purpose connects them.

In Panama, the purpose was not "reduce mosquitoes" as an abstract ecological goal. It was to reduce disease transmission and protect the capacity to live and work. Mosquito counts, breeding sites, cases, deaths, housing, drainage, and response to intervention formed parts of the state.

In a patient, the purpose may be to restore function, prevent irreversible injury, or improve survival - not merely normalize a biomarker.

In a spacecraft, the purpose may be to preserve crew health and mission capability within limited resources - not to maximize microbial killing.

A purpose statement should include the beneficiary, outcome, time horizon, and unacceptable trade-offs.

For example:

"Maintain safe potable water for a six-person crew during a 900-day mission while minimizing consumables, antimicrobial resistance, material damage, and false emergency shutdowns."

That sentence is far more useful than "monitor the microbiome."

The problem of multiple purposes

Real systems serve purposes that conflict.

A hospital water system must limit microbial risk, avoid scalding, conserve energy, preserve materials, and remain usable. A farm must produce food, protect soil, remain economically viable, and manage environmental effects. A medication must improve efficacy, limit toxicity, be manufacturable, affordable, and acceptable to patients. A space habitat must protect crew, conserve mass and power, support science, and remain repairable.

The Universal Control Layer should not hide trade-offs inside one score.

It should show a purpose hierarchy:

1. non-negotiable safety and rights;
2. primary mission outcome;
3. resource constraints;
4. secondary benefits;
5. optimization preferences.

When purposes conflict, authority and escalation rules should be explicit.

In the Canal Zone, the technical mission and the rights of local people were not treated equally. That historical failure belongs in the control analysis, not in a separate moral appendix.

A system that optimizes the mission by externalizing harm has not solved the system. It has redrawn the boundary to hide the harm.

The boundary test

Every control system chooses a boundary.

Fordlandia's boundary centered the plantation and treated the surrounding ecology as context.

Biosphere 2's glass made the material boundary visible, but management, energy, knowledge, and human relationships crossed it.

A company can report lower emissions by moving production to a supplier. A hospital can reduce infection rates on one unit while transferring high-risk patients elsewhere. A drug can improve one endpoint while shifting cost to another organ or population.

The boundary test asks:

- What inputs and outputs cross the declared system?
- Who experiences costs outside the boundary?
- Which delayed effects are excluded?
- Which populations are not represented?
- Does the intervention move the problem rather than reduce it?

Control becomes more honest as the boundary expands enough to include the important consequence.

It cannot include everything. It must include what the purpose makes morally and causally relevant.

Verification: the intervention must answer back

After an intervention, the system must be measured again.

Did yellow fever cases fall?

Did malaria incidence change?

Did the vector population decline in the relevant places?

Did control remain effective through season and migration?

What unintended effects occurred?

The answer should update the model, not merely justify the original decision.

This is the difference between adaptive control and policy defense.

Institutions often become invested in an intervention. Budgets, careers, public promises, and political identities attach to it. Negative feedback is then interpreted as an implementation problem rather than evidence that the model is wrong.

The Universal Control Layer needs independent verification and precommitted failure conditions.

What result would make us stop?

What result would make us change the target?

What result would make us reduce authority?

A controller that cannot revise its own intervention is only a mechanism for persistence.

The least destructive effective action

The intervention ladder can be summarized as one operating principle:

> ***Use the least destructive action that can change the trajectory within the available time.***

"Least destructive" does not mean weakest. It means smallest collateral disruption relative to the purpose.

A targeted vector-control program can be less destructive than treating every infected person after transmission. A narrow-spectrum antimicrobial can be less disruptive than a broad-spectrum one when the pathogen is known. A local gut-restricted therapy can be preferable to systemic exposure when the relevant target is local. A change in airflow can be preferable to repeated surface disinfection. A multi-target drug can be preferable to a maximally selective agent if the disease network requires coordinated modulation and the off-target risks are understood.

The last example leads to the next chapter.

For much of modern drug discovery, selectivity became a mark of quality. A molecule that interacted with multiple targets could be called dirty or promiscuous. Sometimes that judgment protected patients. Unintended interactions cause toxicity.

Sometimes it hid a different possibility.

In the mid-1990s, HIV was defeating one drug after another through mutation and selection. The virus exploited the vulnerability of a single point of attack. The trajectory changed when clinicians combined drugs that acted at different points in the viral life cycle.

The treatment was not one magic bullet.

It was a designed chord.

CHAPTER 8

The Cocktail That Bent the Curve

A rapidly adapting network yielded when medicine stopped relying on one point of attack

---

In the first years of the AIDS epidemic, the curve seemed to point in only one direction.

People became infected with HIV, often without knowing it. The virus attacked the immune system. Opportunistic infections appeared. Cancers emerged. Friends and families watched young adults disappear with a speed that medicine could not stop. Hospitals created entire wards. Activists fought stigma, delay, and indifference while researchers raced to understand a virus capable of copying itself imperfectly enough to generate variation and efficiently enough to overwhelm defense.

Zidovudine - AZT - became the first antiretroviral drug approved in the United States in 1987. It inhibited reverse transcriptase, an enzyme HIV uses to copy its RNA into DNA. For some patients, it produced benefit. It also revealed the limits of one vulnerable intervention. Toxicity constrained dosing. The virus continued to replicate. Resistant variants emerged.

Additional nucleoside analogues followed. They could be combined, but the disease remained difficult to suppress durably.

Then protease inhibitors changed the available architecture.

HIV protease cleaves viral polyproteins into functional pieces required for mature infectious particles. Inhibiting the enzyme attacked a different point in the life cycle. By the mid-1990s, clinicians began combining two nucleoside reverse-transcriptase inhibitors with a protease inhibitor. The regimens were demanding: many pills, strict timing, substantial adverse effects, and complex interactions. Yet the biological logic was powerful.

The virus now had to escape several barriers at once.

Clinical trials in 1997 showed that three-drug combinations could suppress viral load far more effectively than earlier regimens and slow disease progression. Across clinical practice, the population outcome became visible. A landmark analysis reported that mortality among patients with advanced HIV infection in a large U.S. observational cohort fell from 29.4 deaths per 100 person-years in early 1995 to 8.8 by the second quarter of 1997. Major opportunistic infections fell sharply as use of intensive antiretroviral therapy increased.

The curve bent.

The term highly active antiretroviral therapy - HAART - entered medicine. HIV did not disappear. Treatment did not become easy. Access remained profoundly unequal. Resistance, toxicity, adherence, cost, and stigma continued. But for many people with access to effective therapy, a rapidly fatal infection became a manageable chronic condition.

The breakthrough was not one molecule.

It was a control strategy for an adapting network.

The measurement that made the turning point visible

The AIDS crisis did not yield because medicine suddenly discovered the idea of using several drugs. Tuberculosis, cancer, and other diseases had already taught versions of that lesson. The turning point required drugs that attacked different steps, measurements that could reveal what the virus was doing, and clinical evidence strong enough to change practice.

In the early years of HIV treatment, physicians often had to judge progress through symptoms, opportunistic infections, and CD4-cell counts. Direct measurement of viral RNA transformed the control problem. Viral load made replication visible before clinical decline. It showed whether a regimen was suppressing the system, whether rebound was occurring, and how rapidly the virus responded.

The laboratory value was not merely a diagnostic. It became feedback.

Trials reported in the mid-1990s demonstrated the power of combinations that included a protease inhibitor with nucleoside analogues. In one landmark study, a three-drug regimen containing indinavir produced much stronger and more durable viral suppression than a two-drug comparison. In another, adding a protease inhibitor to nucleoside therapy reduced progression to AIDS or death. By the end of the decade, population data showed a steep decline in illness and mortality among people with advanced HIV infection who had access to the new regimens.

The curve bent because the loop became more complete:

- measure the virus;
- attack more than one vulnerable function;
- verify suppression;
- detect rebound;
- change the regimen when resistance or intolerance emerged.

This architecture is now so familiar that its novelty is easy to miss. Before reliable viral-load monitoring, the system could be failing molecularly while appearing stable clinically. Before potent combinations, each drug gave the virus a narrower obstacle to route around. Before organized treatment guidelines and adherence support, a theoretically effective regimen could fail in use.

The medicine was a network. The measurement, delivery system, and patient had to become a network too.

The activists who changed the control room

The scientific story cannot be separated from the people who forced institutions to move.

AIDS activists learned the language of trials, endpoints, regulatory review, access, toxicity, and evidence. Groups such as ACT UP challenged slow processes, exclusionary trial designs, inadequate communication, and the assumption that patients should wait quietly outside the technical conversation. Their methods were confrontational because the cost of delay was measured in funerals.

The result was not the abandonment of rigor. It was a renegotiation of where rigor belonged, how quickly evidence could be generated, who could participate in decisions, and what uncertainty people facing a fatal disease were willing to accept.

This is a direct connection to *Breakthrough Aggressiveness* without repeating its scale. The HIV response changed operating posture because the human cost of delay changed the ethical calculation. It also changed back as therapies became chronic, safer options emerged, and long-term toxicities mattered more. Acceleration and validation were not one permanent setting.

The Universal Control Layer adds another lesson: the governed person must have standing in the loop.

A viral-load threshold may justify a treatment change, but the patient experiences the regimen's nausea, fatigue, stigma, schedule, cost, and interaction with life. A technically optimal control action that cannot be lived is not optimal.

Resistance is probability multiplied by opportunity

Resistance can sound like intention: the virus "figures out" the drug. The underlying process is mutation, selection, population size, replication, and time.

A high viral burden creates many replication events. Mutation produces variation. A drug suppresses susceptible variants more strongly than variants with a useful resistance change. If replication continues in the presence of partial drug pressure, the resistant population gains opportunity.

Combination therapy changes the probability structure. When several drugs act through independent vulnerabilities and all reach effective exposure, a virus may need multiple compatible changes before it can replicate successfully. The exact mathematics vary by drug, mutation pathway, fitness cost, and adherence, but the principle is powerful: make the easiest escape route require more simultaneous luck than the population can usually supply.

This is selective polypharmacology at the level of a regimen.

The system fails when the combination exists on paper but not in the body. Missed doses, drug interactions, malabsorption, unequal half-lives, poor tissue penetration, or interrupted supply can turn a multi-drug barrier into sequential single-drug pressure. The architecture must therefore control exposure as well as selection.

For a designed multi-target molecule, the analogous questions are:

- Are all intended interactions achieved at the same clinically realistic exposure?
- Does one target require a concentration that creates unrelated off-target effects?
- Are the interactions simultaneous, sequential, tissue-specific, or state-dependent?
- Can the system evolve or reroute around the pattern?

- What measurement shows that the intended network state, not merely one binding event, has changed?

A chord is not useful because it contains more notes. It is useful because the notes arrive in the right relationship.

Adherence is part of molecular design

The early combination regimens could be difficult. Patients managed numerous pills, food restrictions, dosing intervals, and side effects. The pharmacology may have been sophisticated while the user experience was punishing.

Every additional burden creates another failure path.

This is why later progress included not only new targets and better potency, but simpler dosing, co-formulated tablets, improved tolerability, longer-acting options, resistance barriers, and better support. The control action became easier to maintain in the world where people actually live.

The same principle should shape future network medicines and remote-habitat interventions. A therapy requiring perfect timing, complex preparation, or continuous expert supervision may be biologically elegant and operationally fragile. A slightly less potent intervention with a wider margin, simpler use, and reliable verification may deliver greater real benefit.

Human factors are part of pharmacodynamics because the drug that is not taken has no target engagement.

The access fracture

Effective combination therapy produced one of medicine's great reversals, but the reversal was distributed unequally.

For years, high prices, patents, manufacturing capacity, health-system weakness, stigma, and political denial limited access across many countries with enormous need. Global activism, generic competition, international financing, procurement programs, and public-health infrastructure changed the trajectory, but not without delay and conflict.

A Universal Control Layer that can diagnose and modulate only for wealthy systems is not universal in any meaningful sense.

Access has to be designed into the architecture:

- Which measurements are available outside advanced centers?
- Can the intervention be manufactured and distributed at scale?
- Does it require cold chain, specialist interpretation, or continuous connectivity?
- Who owns the data and the intellectual property?
- Who bears the burden of false alarms or incomplete models?
- Can the system work with low-resource fallback modes?

The purpose is not achieved when the control action exists. It is achieved when the people for whom it was intended can reliably receive benefit.

The eighth law: adaptive systems require more than one vulnerable point

The eighth law of the Universal Control Layer is:

> ***When the system can adapt around one intervention, control must account for the escape path.***

A single intervention can be appropriate when the target is essential, the effect is durable, the system has little capacity to compensate, and the downside of broader action is high.

It becomes fragile when:

- the target can mutate;
- parallel pathways can compensate;
- the intervention does not fully suppress the process;

- the system contains heterogeneous subpopulations;
- exposure varies across compartments;
- the control action selects for escape.

HIV combined several of these conditions. Its replication produced variation. Incomplete suppression allowed selection. Different drugs had different resistance pathways. A coordinated regimen reduced the probability that one variant could escape all components simultaneously.

The same principle appears in tuberculosis treatment, cancer therapy, obesity treatment, pest management, cybersecurity, and organizational resilience.

The control system should model not only the intended effect but the evolutionary response to the effect.

Combination therapy is not the same as polypharmacology

The language can become confused.

Combination therapy uses two or more separate interventions.

Polypharmacology describes one compound acting at multiple targets, intentionally or unintentionally.

Network pharmacology considers how interventions affect connected biological pathways and phenotypes.

All can produce coordinated multi-node control. Their design, manufacturing, dosing, attribution, and safety challenges differ.

The HIV cocktail used multiple drugs, each with its own pharmacology. The regimen could be adjusted by changing components. Toxicity and interactions could be traced more readily than in a single molecule with many activities. Adherence was difficult because the patient had to manage the combination.

A single multi-target molecule can simplify dosing and create coordinated exposure, but it may be harder to optimize every interaction or separate benefit from toxicity. A molecule that binds ten proteins at a screening concentration is not automatically a rational polypharmacological drug.

The Universal Control Layer needs a stricter vocabulary.

Intended node. A target selected because modulating it contributes to the therapeutic control objective.

Supportive node. An additional target whose partial modulation improves network behavior or limits compensation.

Neutral interaction. A measurable interaction without meaningful effect at therapeutic exposure.

Hazardous off-target. An interaction that contributes unacceptable risk.

Context-only interaction. An effect relevant only in a compartment, concentration, conformation, or state unlikely to occur clinically.

The goal is not "more targets." It is the right pattern at the right exposure.

The promiscuity myth - and the promiscuity reality

Natural products are sometimes dismissed as promiscuous when they interact with multiple proteins in screening panels. The criticism can be fair. Reactive, aggregating, membrane-disruptive, or generally toxic compounds can produce broad assay activity that has little therapeutic value.

The criticism can also be an artifact of method.

A molecule tested at a concentration far above its natural or therapeutic exposure may bind targets it would never meaningfully encounter in vivo. An isolated protein can present surfaces that are buried in cells. A molecule can be active only in one conformational state, tissue, pH, metal environment, or membrane. A family of related targets can be engaged intentionally because they coordinate one physiological function.

Nature does not guarantee elegant specificity. Evolution optimizes fitness in context, not drug development. It produces toxins, detergents, covalent warheads, and broad stress chemistry as well as exquisitely tuned signals.

The right conclusion is not that nature never builds promiscuous molecules.

It is that **assay breadth must be interpreted through exposure, kinetics, location, and function**.

A key can fit several locks. The question is which locks it reaches, at what concentration, for how long, and with what consequence.

A chord, not a note

The musical metaphor helps if we do not let it become decoration.

A note has pitch, duration, volume, and timing. A chord has relationships among notes. The same notes played in another order or register can produce a different effect.

A multi-target drug profile also has relationships:

- affinity at each target;
- efficacy - activation, inhibition, partial agonism, bias;
- exposure in each tissue;
- residence time;
- sequence of engagement;
- metabolite activity;
- target abundance and state;
- interaction with other treatments;
- patient-specific biology.

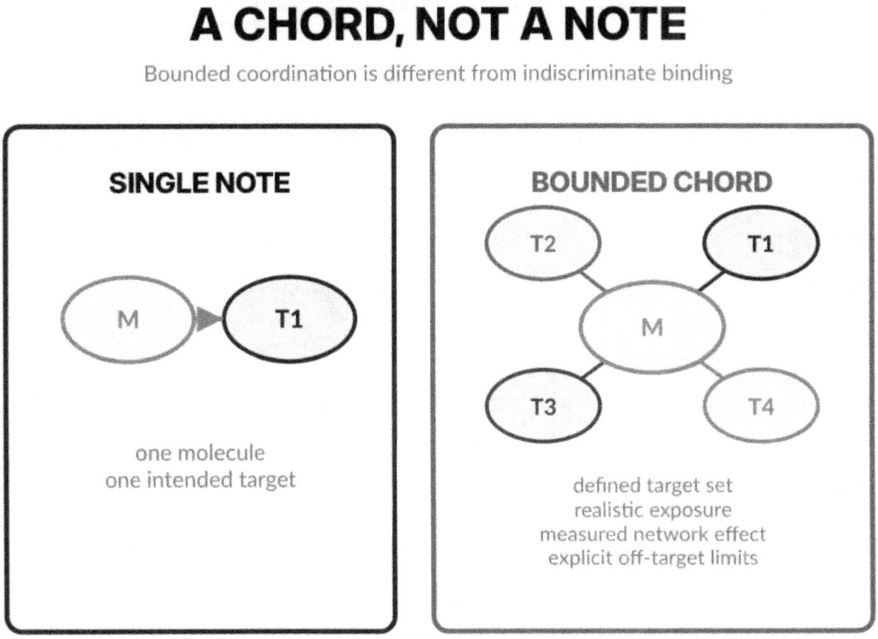

Figure 11. A chord, not a note. Selective network modulation is defined by the pattern of targets, strengths, timing, location, and exposure - not by target count alone.

A useful polypharmacological profile might partially inhibit one node, activate another, and avoid a third. Maximum potency at every target could be worse than calibrated modulation.

This is why de novo multi-target design is difficult. Medicinal chemistry traditionally optimizes one structure-activity relationship while controlling liabilities. Polypharmacology requires simultaneous optimization across several desired and undesired relationships, plus pharmacokinetics and safety.

Artificial intelligence can explore this multidimensional space. It still needs ground truth about which target combination matters and which profiles produce real phenotypes.

The missing data problem returns.

Find the key from the lock - or the lock from the key

Drug discovery can begin in two directions.

Target-first. Identify a protein or pathway implicated in disease, then search for molecules that modulate it.

Phenotype-first. Identify a molecule or fraction that changes a relevant biological state, then determine the targets and mechanism.

Both are valuable.

Target-first discovery provides a clear hypothesis, assay, and optimization path. It can become narrow when the selected target represents only one part of the disease network or when the assay strips away context.

Phenotype-first discovery preserves more system behavior. It can reveal unexpected mechanisms and useful multi-target profiles. It can become difficult when the active material is complex, the target is unknown, or the phenotype does not translate.

The Universal Control Layer connects the two through iterative deconvolution.

If we have the lock, search natural and designed chemistry for keys.

If we have the key, map the locks, exposure, and network response.

Then return to the system and verify that the proposed mechanism predicts the intervention outcome.

The loop prevents either direction from becoming dogma.

Known targets and unknown language

The human proteome contains many targets with incomplete chemical tools. Structural biology, genetics, disease biology, functional genomics, and initiatives to create probes for more proteins are expanding the map. This is essential progress.

The Ghost Genome adds another source of keys.

Microbial and environmental chemistry can contain structures shaped by interactions with proteins, membranes, metals, nucleic acids, and other organisms. Some molecules may directly engage human targets. Others may inspire mimics. Others may reveal scaffolds capable of coordinated multi-target behavior. Still others will be irrelevant or dangerous.

Biosortia's current strategy, as a commercial illustration, uses known and hypothesized therapeutic targets to interrogate natural-product chemistry and computationally generated analogs. This is not a claim that the full Universal Control Layer exists. It is one practical bridge: use the growing map of disease-relevant targets while expanding the chemical priors brought to those targets.

The long-term vision is larger.

As the language becomes better understood, discovery can progress from finding natural keys to designing **complex mimics** and eventually **de novo selective polypharmacological agents** whose profiles are specified for a network state.

That progression requires evidence at each step:

1. identify a reproducible molecule or family;
2. confirm structure and stereochemistry;
3. establish relevant exposure;
4. map intended and unintended targets;
5. show a network phenotype;
6. optimize the profile without losing the phenotype;
7. verify in relevant models;
8. test whether the intervention improves a human outcome.

The language of control becomes valuable only when it survives translation.

Concentration is grammar

A word can change meaning by context. A molecule can change function by concentration.

Hormones act across narrow ranges. Neurotransmitters produce local pulses. Reactive oxygen species can participate in signaling at controlled levels and damage cells at higher levels. Bile acids act as detergents and receptor ligands. Nitric oxide can signal or contribute to toxicity depending on source, amount, and location.

A screening panel often reports whether a compound interacts with a target at one or several concentrations. A control model needs a fuller grammar:

- free concentration at the target;
- peak and trough;
- local versus systemic exposure;
- pulse versus sustained delivery;
- active metabolites;
- competitive endogenous ligands;
- target occupancy and reserve;
- downstream adaptation.

"Hits multiple targets" is not a therapeutic sentence. The grammar supplies the meaning.

This is one reason local delivery can be powerful. A gut-restricted molecule can modulate intestinal targets while limiting systemic exposure. An inhaled drug can reach the lung. A topical agent can act at skin. A tethered antimicrobial can alter a device surface. The same chemistry delivered systemically could produce a different risk-benefit profile.

The Universal Control Layer is spatial.

Resistance as feedback

Resistance is the system answering the intervention.

In HIV, resistant variants showed which parts of the control strategy were vulnerable. In bacteria, resistance mechanisms reveal transport, target modification, degradation, efflux, biofilm protection, and community support. In cancer, relapse can reveal clonal selection and pathway compensation. In agriculture, pest resistance reveals exposure patterns and inherited escape.

A control system should treat resistance data as feedback, not only failure.

That means collecting:

- pre-intervention diversity;
- exposure distribution;
- time to escape;
- genetic and phenotypic mechanisms;
- fitness cost of resistance;
- cross-resistance;
- collateral sensitivity;
- environmental reservoirs;
- response to withdrawal or rotation.

The intervention strategy can then change: combination, sequence, dose, timing, localization, cycling, or a different target.

The goal is not to stay permanently ahead of evolution. It is to design control that anticipates adaptation and preserves options.

The safety problem of coordinated action

Multi-node control can improve efficacy. It can also multiply uncertainty.

Targets participate in several tissues and pathways. A combination that benefits one state may harm another. Partial effects can add or interact nonlinearly. Pharmacokinetic variability can change the intended balance. One component can inhibit the metabolism of another. A multi-target molecule cannot be independently adjusted easily after dosing.

The Universal Control Layer requires a **network safety envelope**.

Instead of asking only whether each target is safe independently, the system asks:

- Which combinations have been observed in humans or relevant models?
- What compensatory functions are being removed together?
- Which tissues receive the combined exposure?
- What biomarkers indicate beneficial engagement?
- What biomarkers indicate approaching toxicity?
- Can the profile be reversed or rescued?
- How does the envelope vary among patients?

This is a natural application for closed-loop monitoring. A therapy can be adjusted using both efficacy and toxicity signals rather than fixed dosing alone. The sophistication must be matched to evidence. An adaptive regimen with poor sensors can be more dangerous than a simple validated one.

Combination is also an organizational principle

The HIV breakthrough required more than molecules. It required virology, medicinal chemistry, clinical trials, regulatory action, manufacturing, activism, patient adherence, diagnostics, public funding, and access systems.

A control layer is itself polyfunctional.

No one discipline can own it.

Chemists understand structure and synthesis. Biologists understand pathways and phenotypes. clinicians understand patients and operational consequence. Engineers understand sensors, control, and failure modes. Data scientists understand inference. Ethicists, regulators, and affected communities understand legitimacy and harm. Manufacturers understand what can be produced reliably.

The teams will have different professional comfort zones. The volume II book addressed how those differences influence operating posture. Here the implication is practical: the control architecture should assign decision rights by consequence rather than prestige.

The person who builds the model should not be the only person who validates it.

The person who discovers the signal should not be the only person who decides it is actionable.

The person who funds the program should not be able to redefine success after the result.

Checks and balances are part of biological control because institutional behavior changes the intervention.

The selective-polypharmacology design brief

A project seeking a multi-target therapeutic can use a one-page design brief.

Control purpose. What disease state or function should change?

Network hypothesis. Which nodes and relationships sustain the state?

Desired profile. Direction and approximate strength at each intended node.

Spatial profile. Where should the molecule act, and where should exposure be minimized?

Temporal profile. Pulse, sustained, conditional, or stage-specific action.

Liability exclusions. Targets, tissues, chemistries, and metabolites to avoid.

State biomarkers. Signals of target engagement, network response, efficacy, and approaching harm.

Escape model. How can the system compensate or resist?

Fallback. What action is taken if the profile is wrong?

Evidence gate. What proof is required before higher-consequence testing?

This brief turns the chord metaphor into an engineering specification.

The curve beyond medicine

Multi-node control applies to habitats too.

A closed-loop water problem may require changing temperature, flow, nutrients, surface condition, disinfectant, and monitoring together. A crop disease may require genetics, planting structure, biological control, and targeted chemistry. Crew cognitive performance may require sleep scheduling, carbon-dioxide control, nutrition, exercise, light, and medical support.

No single actuator should be expected to carry the whole system.

The next chapter considers the opposite side of the control problem: once the system knows what intervention it needs, can it produce that intervention where and when required?

In 1978, a small biotechnology company and its collaborators inserted synthetic human insulin genes into bacteria. Four years later, the FDA approved the first recombinant-DNA-derived medical product.

Biological information had become a manufacturing instruction.

CHAPTER 9

The Message That Became Medicine

Biological information became a manufacturing instruction, opening a path from data to matter

---

The first recombinant human insulin did not begin in a pancreas.

It began as information.

For decades, people with diabetes depended on insulin extracted from animal pancreases. The medicine transformed a fatal disease, but the supply chain required slaughterhouse material, purification, standardization, and management of differences between animal and human proteins. As the number of patients grew, scientists and companies searched for a more scalable source.

Recombinant DNA offered a new route.

In the late 1970s, researchers associated with Genentech synthesized DNA sequences encoding the chains of human insulin and introduced them into *Escherichia coli*. The bacteria produced the components. The chains were purified and assembled into biologically active insulin. Eli Lilly developed the manufacturing process and product. On October 28, 1982, the U.S. Food and Drug Administration approved Humulin, the first biosynthetic human insulin and the first approved medicine produced using recombinant DNA technology.

The achievement changed more than insulin supply.

A therapeutic protein could be represented as sequence, transmitted as information, inserted into a production organism, manufactured under controlled conditions, purified, tested, formulated, and delivered.

The message had become medicine.

This is one of the conceptual bridges to deep-space survival. Physical medicines have mass, volume, shelf life, storage conditions, and limited inventories. Information can encode many possible products at a fraction of the transport burden. A habitat that can translate validated biological instructions into safe local products gains flexibility.

But the phrase "send the code, not the drug" hides most of the work.

The code is only the beginning.

Before the message, the animals

Insulin had been a medicine for six decades before recombinant DNA changed how it was made.

After Frederick Banting, Charles Best, James Collip, and John Macleod helped turn pancreatic extracts into a life-saving treatment in the early 1920s, industrial producers purified insulin from the pancreases of pigs and cattle. The medicine transformed type 1 diabetes from a rapidly fatal disease into a condition people could survive, but the supply chain depended on slaughterhouses, extraction, purification, standardization, and enormous biological variability.

Animal insulin was not a primitive failure. It was one of the great manufacturing achievements of twentieth-century medicine. It also revealed the limits of sourcing a human therapy from a finite stream of animal organs. Purity improved over time, but immune reactions, supply, batch control, and molecular differences remained relevant.

Recombinant production changed the source model.

Scientists inserted genetic instructions for human insulin components into microorganisms, cultivated the organisms, recovered the products, assembled or processed them, purified the result, and proved that the final medicine met identity, potency, purity, and safety requirements. In 1982, the U.S. Food and Drug Administration approved Humulin, the first approved therapeutic made using recombinant DNA technology.

The popular sentence—"put the human insulin gene into bacteria"—contains the breakthrough and hides almost all of the work.

The cell did not become a pharmacy merely because it carried DNA. Expression had to be controlled. The product had to be separated from cellular material. Correct chemical form mattered. Impurities and process contaminants mattered. Analytical standards mattered. Manufacturing reproducibility mattered. Clinical comparison mattered.

Information became medicine only through process.

That distinction protects this book from one of its own seductive metaphors. Biological software is not executable anywhere by anyone with a sequence. Every execution environment has constraints, error modes, materials, energy demands, and quality requirements.

The manufacturing chain is a second biological system

A biomanufacturing process has its own state vector.

Cell strain.

Genetic construct.

Growth phase.

Nutrients.

Temperature.

Oxygen transfer.

pH.

Induction.

Product folding or assembly.

Host-cell proteins.

Endotoxin or other contaminants.

Purification performance.

Storage and delivery.

Changing one variable can move several others. The organism adapts to production stress. A high-yield condition can increase an impurity. A purification step can remove a contaminant while damaging potency. A process that works at milliliter scale can fail when mixing and oxygen transfer change in a large vessel.

This is another closed loop: measure the production state, adjust the process, verify the product, and learn from deviation.

The Universal Control Layer should not collapse the patient loop and the manufacturing loop into one model. They are connected but governed by different evidence. The process must release a product before the product is allowed to influence the patient. The patient's response then creates post-use evidence that may change manufacturing, labeling, dosing, or selection.

Local production adds a third loop: the mission environment. Power, consumables, vibration, operator skill, contamination control, waste, and storage all influence whether a validated Earth process remains qualified at the edge.

The data payload therefore contains nested control systems, not a recipe card.

Small molecules and biological products require different factories

The idea of "printing a medicine" can obscure the diversity of what medicines are.

A small organic molecule may require several chemical steps, catalysts, solvents, purification, stereochemical control, and crystallization. A peptide may be made chemically or biologically, then folded or cyclized. A protein may require a living or cell-free expression system and careful control of structure and modifications. A bacteriophage, cell therapy, RNA product, or microbial consortium creates still other manufacturing and release problems.

No single onboard printer will make all of them.

A realistic mission architecture would begin with a bounded formulary selected for:

- high clinical value relative to stored mass;
- plausible demand during the mission;
- stability of digital and physical inputs;
- manufacturing simplicity;
- available analytical release tests;
- manageable waste and contamination risk;
- safe fallback if local production fails.

The first locally produced products may be nutrients, diagnostic reagents, simple proteins, or narrowly defined therapeutic components—not an improvised cure for any unknown disease.

The value of the long-term vision is not that every product will be synthesized on demand. It is that the boundary between what must be carried as finished inventory and what can be carried as validated production capacity can be moved deliberately.

The terrestrial frontier comes first

Remote medicine on Earth already presents versions of the space problem.

A rural clinic may lack cold-chain reliability, specialist laboratories, or rapid resupply. A military field hospital may face uncertain demand and damaged logistics. An outbreak region may need diagnostic reagents faster than centralized manufacturing can deliver. A disaster can interrupt transport while changing the disease and injury profile overnight.

These settings offer a more accessible proving ground for portable production and ISMM-informed diagnostics than a first deployment on Mars.

They also expose equity and governance questions early. A portable system that works only with proprietary cloud access or rare consumables may shift dependency rather than remove it. A local unit that can make a high-value product but cannot verify identity or contamination creates dangerous confidence. A system that sends patient chemistry to a remote model without meaningful consent may solve logistics by creating surveillance issues.

Earth-first development should therefore test the entire service, not only the manufacturing reaction:

- training and workflow;
- local quality assurance;
- supply of cartridges and reagents;
- cybersecurity;
- regulatory responsibility;
- waste handling;
- product recall;
- maintenance;
- patient follow-up;
- economic access.

Space is the extreme use case. Responsible terrestrial deployment is the apprenticeship.

Who is allowed to update the recipe?

Digital manufacturing creates a new form of change control.

In a conventional factory, an approved process change passes through documented review, validation, and regulatory oversight. In a distributed network, the product definition or process file could be updated remotely. That creates speed and also a new attack surface.

A future system may need to distinguish several update classes.

Administrative update. Documentation or interface changes with no effect on product quality.

Analytical update. Improved interpretation of sensor or quality-control data within an already validated process.

Process-bound update. A change allowed inside a prevalidated design space, such as a controlled adjustment for a known environmental condition.

Product update. A change to the molecular product, dose, formulation, or indication requiring a much higher evidence and authority threshold.

Emergency experimental update. A last-resort action under conditions where no validated option exists and the cost of inaction is extreme.

The control layer must know which kind of update it is making. A software culture accustomed to continuous deployment cannot be imported without translation into a system where one altered parameter can change the medicine.

Version control is necessary.

Biological consequence control is the real requirement.

The ninth law: information becomes capability only through a qualified process

The ninth law of the Universal Control Layer is:

> ***A biological instruction is not a product until the process can make, verify, and control it.***

A DNA sequence does not contain fermentation conditions, expression yield, folding, post-translational modifications, purification, contaminants, potency, stability, dose, delivery, or clinical context. Two systems can execute the same sequence differently. A protein can be present and inactive. A chemically correct product can be impure. A pure product can be delivered at the wrong dose. A validated product can degrade in storage.

The Universal Control Layer must connect three forms of knowledge.

Design knowledge. What molecular intervention is intended?

Process knowledge. How can it be made reproducibly with available equipment and inputs?

Use knowledge. How should identity, purity, potency, dose, route, timing, and response be verified?

The complete payload is not one sequence. It is a controlled recipe and evidence package.

From factory to production unit

Recombinant insulin was made in centralized industrial facilities because scale, quality assurance, and regulatory control favored centralization. That model remains essential for many medicines.

Remote environments create a different optimization problem.

A Mars mission cannot stock every product that might be needed over several years. Some medicines degrade. Some have uncertain demand. Resupply and emergency return are limited or impossible. A crew may encounter a pathogen, injury, nutrient deficiency, or individual response not covered by the original manifest.

Local biomanufacturing becomes attractive when the mass and flexibility advantages exceed the complexity and risk.

Research systems have begun exploring parts of this possibility. Freeze-dried cell-free transcription and translation machinery can be stored and rehydrated to produce proteins from DNA templates. In 2016, Keith Pardee and colleagues demonstrated portable, on-demand biomolecular manufacturing using freeze-dried cell-free systems. NASA's BioNutrients experiments have tested engineered microorganisms that produce nutrients on demand aboard the International Space Station. Other NASA-supported programs examine biological production of food, materials, and potentially pharmaceuticals for long-duration missions.

These efforts are early relative to an autonomous space pharmacy. They show that the boundary between stored product and stored production capacity is moving.

The production ladder

Local manufacture should advance through a ladder of increasing consequence.

Level 1 - Nonmedical indicators and research reagents. Produce low-consequence molecules to validate equipment and process control.

Level 2 - Nutrients and food components. Manufacture compounds with defined quality needs and manageable risk.

Level 3 - Environmental and industrial enzymes. Produce tools for waste processing, materials, or diagnostics.

Level 4 - Diagnostic reagents. Manufacture components whose errors can influence decisions but are not administered directly.

Level 5 - Topical or local low-risk products. Produce interventions with limited systemic exposure.

Level 6 - Established systemic medicines. Manufacture validated products using locked, qualified processes and strong testing.

Level 7 - Mission-specific novel therapeutics. Design or select a new intervention in response to an emerging state and manufacture it locally.

The last level is the science-fiction horizon. It requires advances not only in synthesis but in diagnosis, target selection, toxicology, model confidence, regulatory governance, and monitoring. A system that can manufacture a novel molecule faster than it can establish safety has accelerated the wrong part of the loop.

The ladder allows capability to mature without pretending the final step already exists.

The data payload

A deep-space medical payload can be thought of in four layers.

Physical inventory. Essential medicines, consumables, reagents, and emergency supplies carried from Earth.

Production platform. Hardware capable of making defined classes of products: microbial fermenters, plant systems, cell-free systems, chemical synthesis, purification, formulation, and sterile handling.

Digital library. Validated sequences, structures, process recipes, analytical standards, dosing guidance, contraindications, and contingency plans.

Adaptive design capability. Models that can interpret a new need, select among validated options, and, eventually, propose bounded modifications.

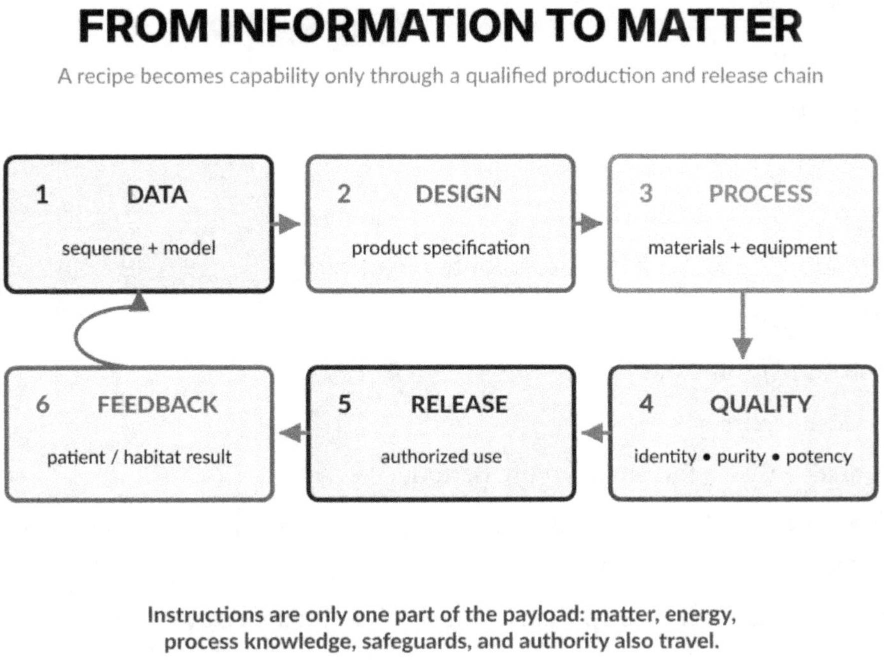

Figure 12. From biological information to local production. A validated design enters a qualified manufacturing process, followed by purification, identity and potency testing, formulation, dosing, response monitoring, and feedback.

The architecture is hybrid. It does not eliminate physical stock. Emergency drugs must remain available even if production fails. The system uses local manufacture for replenishment, short-shelf-life products, specialized needs, and selected contingencies.

Information creates optionality. Physical inventory creates immediate reliability.

Quality control at the edge

A local production system must perform tests that centralized manufacturers conduct with specialized laboratories.

At minimum, it may need to verify:

- identity;
- concentration;
- purity;
- potency;
- sterility or microbial limits;
- endotoxin and other process contaminants;
- aggregation or degradation;
- formulation conditions;
- dose accuracy;
- stability over the intended use period.

The tests must be compatible with mass, power, consumables, crew time, and limited expertise. They need controls and calibration standards. They need to detect when the process is outside its validated range.

This requirement brings the Ghost Genome back into the story in an unexpected way.

Biomanufacturing systems have their own hidden chemistry. Host organisms produce metabolites and impurities. Media components vary. Stress changes output. Contaminants can transform the product. A cell-free system contains molecular machinery extracted from cells. The process state matters as much as the design sequence.

The production unit therefore needs its own Universal Control Layer.

It senses growth, expression, metabolites, impurities, temperature, pH, oxygen, and product quality. It adjusts conditions. It stops when the state becomes unsafe. It records the full batch history.

The pharmacy is a habitat inside the habitat.

What can be digitized - and what cannot

Biological design has digital representations: nucleotide sequence, amino-acid sequence, chemical structure, process parameters, analytical fingerprints. These representations can travel as data.

The molecules themselves remain physical.

Stereochemistry, folding, solvent, crystal form, impurities, particle size, container interaction, and temperature can alter function. A small molecule represented by a structure string still requires a synthetic route, reagents, purification, and confirmation. A peptide sequence can produce several physical forms. A biologic may require modifications not supplied by a simple bacterial host.

The phrase *digitizing biology* should therefore mean creating representations sufficiently faithful to guide physical reconstruction and prediction - not converting life into pure code.

Matter retains veto power.

From known products to adaptive design

The safest local-production model uses an approved or otherwise validated product and a locked recipe. The design is known; only the manufacturing location changes.

Adaptive design is harder.

Suppose a habitat detects an unfamiliar resistant bacterium. The control system might:

1. confirm the organism and phenotype;
2. compare it with Earth reference data;
3. identify known susceptibility and resistance patterns;
4. select an existing validated antimicrobial or combination;
5. manufacture or formulate the product locally if needed;

6. monitor response and toxicity.

That scenario is plausible before truly de novo therapy.

A later system might select a natural-product-inspired molecule or a known scaffold adapted to the target. Later still, it might design a selective multi-target agent based on the observed network and manufacture it. Every step away from known products increases uncertainty.

The controller should expose the distance from precedent.

A **Class A** intervention reproduces a validated product and process.

A **Class B** intervention uses a validated product with a modified formulation or dose.

A **Class C** intervention uses a known scaffold against a new but related target state.

A **Class D** intervention is a novel molecule or profile.

Authority, testing, and acceptable emergency conditions should differ by class.

The Earth-first obligation

The technology required for space can improve remote care on Earth.

Portable diagnostics, room-temperature-stable reagents, on-demand nutrient production, compact purification, and local manufacturing can serve disaster zones, rural communities, military medicine, outbreaks, and supply-chain disruptions. Deep-space constraints force attention to reliability, autonomy, and resource efficiency that wealthy terrestrial systems often avoid.

The reverse is also true. Earth is the ethical and scientific proving ground.

A system should not be sent to Mars with claims that have not survived diverse use on or near Earth. Remote manufacturing must be validated across operators, environments, failures, and populations. Monitoring models must be tested for bias and drift. Security must be challenged. Emergency procedures must be practiced.

The off-world imperative is not an excuse to skip validation.

It is a reason to begin early.

The security problem

A digital therapeutic library and local manufacturing platform create dual-use and cybersecurity risks.

A malicious actor could alter a sequence, process parameter, dose, or identity standard. A compromised model could recommend an unsafe product. A manufacturing system capable of making beneficial molecules could potentially make harmful ones. Intellectual property and access controls could conflict with mission survival.

The Universal Control Layer needs security designed into the loop.

- cryptographic verification of designs and recipes;
- independent checks of critical parameters;
- hardware limits on prohibited operations;
- separation of design, authorization, and production roles;
- audit logs that cannot be altered locally without detection;
- safe libraries available during communication loss;
- emergency authority with post-event review;
- physical testing that verifies the product rather than trusting the file.

The last point is decisive.

The molecule does not care whether the digital signature was valid. The product still must be measured.

The completeness problem

The early draft of this book used a tempting phrase: we must map the complete chemical language before we leave Earth.

Completeness is the wrong standard if it implies a finished dictionary.

Biological chemistry will continue to vary across people, environments, evolution, and time. New states will emerge. Unknowns will remain. A control system waiting for total knowledge will never operate.

The operational goal is **sufficient coverage with known uncertainty**.

We need enough of the language to:

- recognize high-consequence states;
- distinguish common benign variation;
- select validated interventions for likely scenarios;
- identify when the system is outside its knowledge;
- collect the data needed to learn;
- preserve safe fallback when interpretation fails.

The dictionary can be finite in any defined sampling frame and still be enormous. The actionable subset can be much smaller than the total chemistry.

The Universal Control Layer should prioritize vocabulary by decision value.

Which unknowns sit closest to irreversible harm?

Which signals could change intervention early?

Which molecular families provide broad therapeutic option value?

Which habitat states threaten multiple systems?

Which gaps produce the largest false confidence?

This is a portfolio problem, not a census alone.

The local-manufacturing charter

Before a product is approved for local manufacture, the system should document:

Purpose. What mission need does local production solve?

Product class. Nutrient, diagnostic, enzyme, topical, systemic medicine, or experimental agent.

Design authority. Who can release or modify the digital instruction?

Process envelope. Acceptable inputs, conditions, equipment, and operator steps.

Quality tests. Identity, purity, potency, contaminants, dose, and stability.

Failure modes. Contamination, wrong sequence, process drift, sensor error, purification failure, cyberattack.

Fallback. Stored product, alternative therapy, Earth consultation, or mission abort criteria.

Response monitoring. How benefit and harm will be detected after use.

Learning return. Which data and samples will update the Earth reference engine.

The charter turns a futuristic promise into a governable capability.

The person receiving the product

The manufacturing story can distract from the patient.

A locally produced medicine still enters a person whose physiology may be altered by microgravity, radiation, diet, sleep, stress, and other medications. Pharmacokinetics can differ. Immune response can differ. Kidney and liver function can change. The microbial community can transform the drug.

The Universal Control Layer must close the loop after administration.

The product's analytical release is not the final verification. The person's response is.

This is where state monitoring, network design, and manufacturing converge. The system senses a need, selects an intervention, makes or retrieves it, verifies the product, administers it, measures the response, and adjusts.

That is the architecture promised by the title.

It is also the architecture that makes error consequential. Every connection must be bounded.

The last human factor

A production system can be technically flawless and operationally unusable. Crew members may be exhausted, ill, in conflict, or working under emergency load. Instructions that are clear in a laboratory can fail in a cramped habitat. Protective equipment can impair dexterity. A test requiring twelve careful steps may be unrealistic during a crisis.

Human factors are part of product quality.

The same principle appeared in Apollo 7. The command and service module performed remarkably well. The crew completed a demanding test mission that restored confidence after the Apollo 1 fire. Yet a common cold, sleep disruption, schedule pressure, and disagreement over command produced a mission remembered almost as much for conflict as for technical success.

The flight plan treated the crew as the operator.

The illness revealed the crew as part of the loop.

PART IV

CARRY THE LOOP WITH US

CHAPTER 10

The Cold That Rewrote the Flight Plan

A technically successful mission revealed that the human being is not outside the control system

---

Five hours after launch, Wally Schirra reported that he had a cold.

The timing could hardly have been worse.

Apollo 7 lifted from Cape Kennedy on October 11, 1968. It was NASA's first crewed flight after the Apollo 1 fire had killed Gus Grissom, Ed White, and Roger Chaffee during a launch-pad test twenty-one months earlier. The redesigned command and service module had to prove itself. The mission would test propulsion, navigation, environmental control, rendezvous procedures, communications, and the systems required before a crew could be sent toward the Moon.

Schirra, Donn Eisele, and Walt Cunningham understood that more than one spacecraft was on trial. The credibility of the Apollo program was riding with them.

Schirra had flown Mercury and Gemini. He was experienced, exacting, and already expected to retire after the mission. He had spent long periods with the spacecraft during its construction, determined that the crew would not accept assumptions after the Apollo 1 disaster.

Then the commander became sick.

In microgravity, congestion is miserable. Fluid shifts toward the head. Mucus does not drain normally. Sinus pressure, headache, fatigue, and poor sleep accumulate. The cold spread to Eisele and Cunningham. The three men still had an overloaded schedule of tests.

Conflict with Mission Control intensified.

The crew delayed a planned television broadcast until critical rendezvous work was complete. They objected to procedures passed up from the ground. Schirra used his authority as commander to reject tasks he believed interfered with mission objectives. Exchanges became sharp enough that Apollo 7 was later described in popular accounts as a near mutiny.

The most serious disagreement concerned reentry helmets.

The crew feared that wearing helmets would prevent them from pinching their noses and equalizing pressure in congested ears during descent. Ground controllers wanted the helmets on for safety. Schirra refused. The crew reentered without them.

Apollo 7 splashed down safely on October 22 after nearly eleven days in orbit.

Technically, the mission was a triumph. The propulsion system worked. The spacecraft performed better than many had dared hope. NASA gained the confidence to send Apollo 8 around the Moon two months later.

Humanly, the flight exposed a control gap.

The schedule, spacecraft, ground organization, and crew were operating as though the human state were a secondary variable.

A rhinovirus changed the system.

The fire still in the room

Apollo 7 cannot be understood without the three men who never launched.

On January 27, 1967, Gus Grissom, Ed White, and Roger Chaffee died when a fire swept through the Apollo 1 command module during a ground test. The cabin atmosphere, flammable materials, wiring, hatch design, procedures, and organizational assumptions combined into a catastrophe. NASA stopped crewed Apollo flights and redesigned the spacecraft.

Schirra entered Apollo 7 with that history close at hand. He had known Grissom. He had watched a program driven by national urgency discover that confidence and schedule could not substitute for system safety. His insistence on crew authority and spacecraft readiness was therefore not simply temperament. It was partly a response to an organization that had already proved capable of normalizing danger.

Mission Control carried the same scar from the other side. The ground organization needed a disciplined flight that demonstrated the redesigned vehicle was ready. The United States was still trying to meet the end-of-decade lunar goal. Every delayed test created pressure on the missions that followed.

Both sides could claim responsibility.

Both sides could interpret the other's behavior as threat.

That is how control conflict develops inside high-consequence systems. The same evidence is filtered through different histories of failure.

A future control layer that monitors human state but ignores institutional memory will misunderstand the operator. Stress is not only cortisol or heart rate. It is also what the last disaster taught the group to fear.

When the schedule becomes an actuator

A flight plan appears to describe work. In practice, it also acts on the crew.

The timing of sleep, meals, experiments, communication, exercise, medication, and decision points changes physiological state. An overloaded schedule can create fatigue; fatigue can increase error; error can produce more corrective work; corrective work can further reduce sleep. The plan becomes part of the feedback loop.

This is true far beyond spaceflight.

A hospital staffing schedule changes infection risk, attention, and decision quality. A manufacturing quota changes maintenance behavior. A grant deadline changes what evidence gets generated. A quarterly target changes which warning signs a company allows itself to see.

The Universal Control Layer should therefore treat policies and schedules as actuators. They are interventions applied to human and organizational systems.

Before a high-demand plan is imposed, the controller should ask:

- What physiological and cognitive state does this schedule assume?
- What recovery capacity remains?
- Which tasks require peak judgment?
- Which tasks can be deferred, simplified, automated, or moved?
- What signal triggers workload reduction?
- Who has authority to declare that the plan no longer fits the person?

Apollo 7 had no continuous integrated model that could translate congestion, sleep loss, workload, mission phase, and crew conflict into a revised operating posture. Ground and crew negotiated through voice, authority, and frustration.

The mission succeeded. The method does not scale comfortably to a habitat months from Earth.

Monitoring the operator without turning the operator into a prisoner

As biosensors become more capable, it will be tempting to measure everything: sleep, heart rhythm, glucose, hormones, voice, facial expression, movement, cognition, medication use, microbiome, and molecular fingerprints.

The technical rationale is clear. Earlier state detection could prevent errors, illness, conflict, or mission failure.

The governance risk is equally clear. The same data can become a tool of coercion, performance surveillance, discrimination, or retrospective blame. A crew member may hide symptoms if disclosure threatens flight status. An employee may learn to game a fatigue score. A commander may overtrust a model that labels dissent as biological instability.

A purpose-bound system needs explicit rights.

Right to know. People should know what is measured, why, how long it is retained, and who can see it.

Right to contest. A model's interpretation should be challengeable, especially when it changes authority, duty, treatment, or career.

Right to minimum necessary use. Data gathered for medical safety should not automatically become a general performance score.

Right to protected disclosure. Reporting symptoms or uncertainty should not be treated as disloyalty.

Right to human review. High-consequence decisions should not rest on an unexplained model output when qualified human judgment is available.

Reciprocal duty. Crew members also have a duty to disclose conditions that create material risk for others.

Autonomy and collective safety are not opposites. They are design constraints that must be reconciled openly.

The team has a physiology

A team is not an organism in the literal sense, but it develops measurable states.

Who speaks first?

Whose warnings receive follow-up?

How quickly does bad news travel?

Does disagreement close with a decision or continue underground?

Are people sleeping?

Is workload concentrated in the same individuals?

Are errors increasing at interfaces?

Does the group still share the same picture of purpose?

These variables can drift before performance collapses. They resemble the chemical early-warning problem in a person: no single value proves failure, but the trajectory can reveal loss of resilience.

The control action may be rest, schedule change, role clarification, mediation, additional expertise, a technical redesign, or a shift in command authority. It should not default to treating the most visible dissenter as the defective component.

A mature team monitor looks for system conditions that make conflict more likely and truth harder to hear.

This is where the trilogy's second book becomes operational rather than repetitive. Breakthrough Aggressiveness described the need to change operating intensity as the work changes. The Universal Control Layer asks what evidence should trigger that migration when the workers themselves are changing state.

Authority under communication delay

Near Earth, Mission Control can dominate because expertise and resources sit on the ground and communication is nearly immediate. At Mars, a message can take many minutes one way. A crew cannot wait for a conversational loop during fire, decompression, sepsis, toxic exposure, or rapid environmental failure.

Authority must migrate outward.

That does not mean Earth becomes irrelevant. It means the mission is preauthorized to act within defined envelopes, with escalation rules for unfamiliar states. The local system needs enough information, tools, and legitimacy to make a consequential decision while knowing when it has left validated territory.

A useful authority design distinguishes:

- routine autonomous control;
- crew-directed action within validated bounds;
- commander emergency authority;
- distributed consultation when time permits;
- protected dissent and second-check requirements;
- post-action recording and review;
- conditions requiring mission-wide or Earth-side approval.

The difficult cases will not be purely technical. A life-support intervention may consume a resource needed later. Treating one crew member may increase risk to the group. Quarantining a module may jeopardize food production. Returning early may sacrifice years of work and enormous public investment.

No algorithm can remove the moral decision. Good architecture can prevent the decision from being made blind.

The tenth law: the operator is inside the loop

The tenth law of the Universal Control Layer is:

> ***The human decision-maker is a biological subsystem, not an external source of perfect commands.***

Traditional control diagrams place the operator outside the plant. The operator observes, decides, and acts. In a real mission, the operator sleeps, breathes, becomes infected, experiences pain, adapts to carbon dioxide, responds to stress, miscommunicates, and forms relationships with other operators.

The person changes the system and is changed by it.

This does not reduce judgment to chemistry. Schirra's decisions had technical, cultural, and personal causes. Some of his resistance may have protected mission priorities. Some exchanges were unnecessarily confrontational. NASA's ground teams also operated under pressure after a fatal accident and a national deadline.

The lesson is not that a cold "caused a mutiny."

The lesson is that biological state changed communication, tolerance, workload, risk perception, and the ability of the formal plan to fit the moment.

A control architecture that ignores the operator's state will misread behavior as attitude until the system breaks.

The human control stack

A human operator participates through several coupled layers.

Physical capacity. Strength, vision, hearing, dexterity, cardiopulmonary function, pain, hydration, and fatigue.

Metabolic state. Glucose availability, nutrition, hormones, temperature, and energy.

Immune and inflammatory state. Infection, allergy, inflammation, medication, and recovery.

Neural and cognitive state. attention, memory, reaction time, mood, sleep, and decision load.

Social state. Trust, conflict, authority, communication, isolation, and perceived fairness.

Mission state. Workload, time pressure, consequence, novelty, and available support.

These layers cannot be reduced to one wellness score. They can be monitored selectively for defined purposes.

A long-duration mission might track sleep, cognitive performance, carbon-dioxide exposure, symptoms, inflammatory markers, nutrition, hydration, medication, workload, and team communication. The purpose is not surveillance of personality. It is to preserve capacity and detect when the operating plan should change.

Breakthrough Aggressiveness becomes mission posture

The second book in the trilogy described dynamic migration: changing the operating intensity as uncertainty, maturity, reversibility, and urgency change.

Apollo 7 shows what migration looks like inside a mission.

After launch, the program needed high operational discipline. The spacecraft was new, the tests mattered, and errors could be fatal. Ground control carried procedures and system knowledge. The commander carried direct situational awareness and legal responsibility for the vehicle.

When illness and workload changed the crew state, the mission posture should have been reviewed explicitly.

Which objectives remained mission-critical?

Which could be delayed?

Which communication rules were producing friction rather than safety?

What medical state required schedule adjustment?

What authority belonged to the ground, and what belonged to the commander?

The right answer is not simply "listen to the crew" or "follow Mission Control." It is a predesigned migration protocol that uses state evidence and clarifies decision rights.

A mature Universal Control Layer would connect biological state to operational posture without allowing an algorithm to remove human command.

For example, sustained sleep loss, rising carbon dioxide, degraded cognitive testing, and infection could trigger:

- reduced noncritical workload;
- increased cross-checks for high-consequence actions;
- delayed discretionary experiments;
- protected sleep periods;
- medical consultation;
- different communication cadence;
- temporary transfer of tasks;
- heightened monitoring after medication.

The control action changes the mission, not only the person.

The electrical system is chemical

Apollo 7's cold returns us to the most basic claim in the book.

Human thought depends on electrical signaling. Electrical signaling depends on molecular movement and maintained gradients.

Sodium, potassium, calcium, chloride, protons, neurotransmitters, receptors, channels, pumps, ATP, membranes, and supporting cells create the physical conditions of neural activity. Inflammation changes those conditions. Sleep deprivation changes them. Carbon dioxide and pH can change them. Medicines change them. Microbial metabolites can influence some of the pathways around them.

Stop the movement of small molecules and ions, and even the electrical signal ends.

This does not make every decision a microbiome event. It makes biology inseparable from performance.

The Universal Control Layer treats chemical state as a foundational layer of mission capability, alongside mechanical, electrical, computational, and social systems.

The control room moves outward

Near Earth, a crew can rely heavily on ground support. Beyond Earth orbit, communication delay and disruption shift more authority into the habitat.

A Mars crew may wait many minutes for a one-way message. A real-time medical emergency cannot be managed as a conversation with Houston. The habitat must carry knowledge, diagnostic capacity, production options, and decision protocols.

The Universal Control Layer becomes more autonomous with distance.

It should be able to:

- maintain personal and habitat baselines;
- detect high-consequence drift;
- distinguish sensor error from biological change;
- recommend confirmatory tests;
- isolate subsystems;

- select among validated interventions;
- manage scarce inventory;
- support local production;
- monitor response;
- communicate uncertainty to the crew;
- package evidence for delayed Earth review.

Autonomy does not mean the model becomes captain.

It means the crew gains a better instrument for command.

The architecture of authority

A control layer that can influence medicine, food, air, water, and habitat conditions needs explicit authority rules.

Who can authorize a medication?

Who can isolate a water loop?

When can the system act automatically?

What if the commander and medical officer disagree?

What if Earth guidance arrives after local action?

What if the model recommends an intervention outside its validated envelope?

A practical authority architecture uses consequence bands.

Routine automation. Low-consequence actions within validated limits, such as adjusting sampling cadence or maintaining temperature.

Advisory action. Recommendations requiring human approval, such as changing diet, workload, or diagnostic tests.

Protected intervention. Medical or ecological actions requiring two-person verification and defined evidence.

Emergency authority. Preauthorized actions when delay creates immediate danger.

Prohibited action. Interventions the local system cannot perform regardless of model confidence.

Every action should be logged with the state, model version, evidence, authority, and outcome.

The loop must remember its own decisions.

Control without coercion

The word control becomes most dangerous when applied to people.

A system that monitors chemistry, cognition, behavior, and performance could protect a crew. It could also become a mechanism of coercion. Employers, militaries, insurers, governments, and platforms could use biological predictions to classify, exclude, pressure, or punish.

The off-world environment does not erase rights. It makes power more concentrated.

The Universal Control Layer requires a human covenant:

- collect only data tied to defined safety and health purposes;
- separate medical confidentiality from command information where possible;
- disclose what is measured and how it is used;
- allow human challenge of model interpretation;
- prevent nonmedical use without explicit governance;
- distinguish temporary performance support from permanent labeling;
- preserve dignity during illness and conflict;
- audit differential effects across crew members and populations.

A system that keeps the habitat alive by treating people as replaceable components has failed its purpose.

The mission-control review

A team can use a short review when human or habitat state changes.

What changed? Direct observations, not character judgments.

What is the likely time scale? Minutes, hours, days, or persistent.

What mission functions depend on the affected state? Decision, dexterity, communication, immune defense, production, water, air.

What reversible adjustments are available? Workload, schedule, isolation, environment, treatment, redundancy.

What must remain protected? Safety, authority, rights, critical objectives.

What evidence will show recovery or deterioration? Define before acting.

When will the posture be reviewed again? Avoid indefinite emergency mode.

This is the human equivalent of the intervention ladder.

The universality is in the loop

At this point, the book has moved across systems that seem unrelated.

A glass biosphere.

A rubber plantation.

A Philadelphia hotel.

A newborn blood spot.

Two astronaut twins.

Microbes behind orbital panels.

A canal through Panama.

An HIV treatment regimen.

Recombinant insulin and local biomanufacturing.

A congested crew in Earth orbit.

The universal element is not the organism, technology, or scale.

It is the loop:

- define purpose;
- observe state;
- interpret change;
- select a bounded intervention;
- verify response;
- preserve fallback;
- learn.

This loop can fail at every step.

The purpose can be wrong.

The state can be incomplete.

The sensor can be biased.

The interpretation can confuse association with cause.

The intervention can be too broad.

The verification can measure the wrong endpoint.

The institution can refuse to learn.

The Universal Control Layer is not a claim that these problems disappear. It is a way to make them visible before authority and automation magnify them.

The trilogy closes its loop

The three books form a progression, not three versions of the same argument.

The Ghost Genome is the map problem. It asks why biology's chemical execution layer remains so poorly counted and why natural, context-generated molecules can expand discovery.

Breakthrough Aggressiveness is the operating problem. It asks how leaders match speed, rigor, risk, and proof to the work and migrate as the work matures.

The Universal Control Layer is the system problem. It asks how an expanding map and an adaptive operating method become a durable capability to protect and improve living systems.

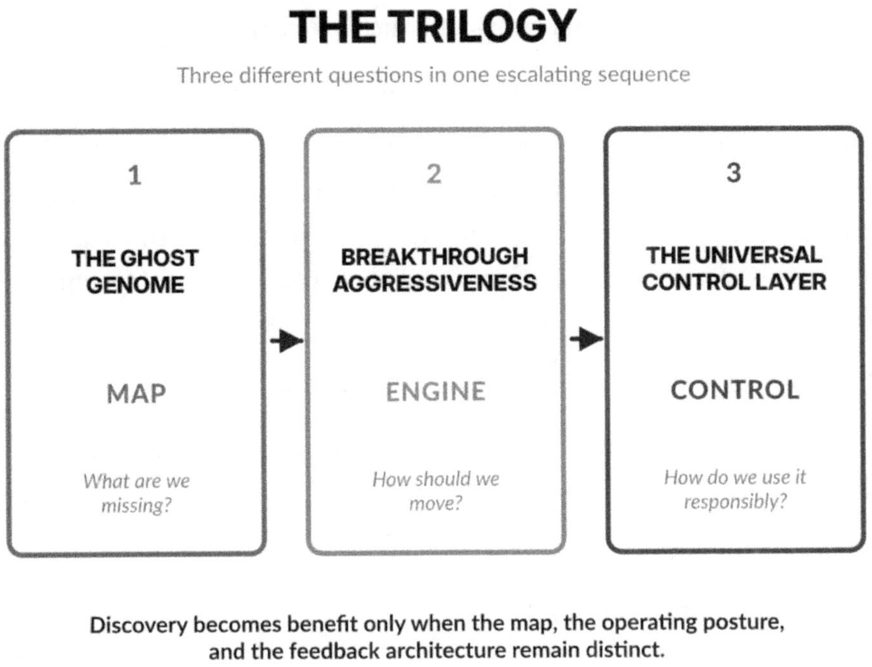

Figure 13. Map, engine, control system. The books are sequential: discover the missing language, organize the work to advance it, and build feedback architectures capable of using it responsibly.

The trilogy does not end with a product.

It ends with an obligation.

The more deeply we can read biological state, the more carefully we must govern the power to change it.

Apollo 7's final answer

The popular memory of Apollo 7 often lingers on the arguments. The mission's final answer was technical success.

The command and service module was ready.

Apollo 8 could go to the Moon.

The crew's conflict did not erase the achievement. The achievement did not erase the conflict. Both belonged to the state of the mission.

That is the mature systems view: do not force the event into one moral.

Schirra was not simply insubordinate. Mission Control was not simply bureaucratic. Illness was not the only cause. The plan was not meaningless. A new spacecraft, a wounded organization, a compressed national goal, an experienced commander, two first-time astronauts, a demanding schedule, and three head colds interacted inside a small capsule.

The system succeeded and still revealed a design flaw.

The next mission inherited the success. A future control layer should inherit the lesson.

The person is not cargo.

The habitat is not background.

The chemistry is not noise.

The model is not the territory.

And the controller is never outside the system it hopes to control.

EPILOGUE

The Smallest Planet We Have Ever Built

The view from Apollo 8 changed the scale of the problem without making the problem simple

---

On December 24, 1968, Apollo 8 emerged from behind the Moon and found the Earth waiting above a gray horizon.

The spacecraft had disappeared from radio contact as it passed around the far side. Frank Borman, Jim Lovell, and Bill Anders were farther from home than human beings had ever traveled. Their mission had been assembled under extraordinary pressure. Apollo 7 had proved the redesigned command module only two months earlier. The lunar module was not ready for the originally planned Earth-orbit test. NASA chose a far more audacious alternative: send the next crew around the Moon.

As the capsule rotated, a blue-and-white sphere entered the window.

Anders reached for a camera.

The photograph that became known as *Earthrise* was not the first image of Earth from space. It became one of the most consequential because the planet appeared neither infinite nor abstract. It was a finite object suspended in darkness, visibly covered by one atmosphere, one ocean system, one moving skin of weather, and no apparent rescue vehicle.

The image invited a sentence that has followed environmental thought ever since: we live on a spaceship.

The metaphor is useful and incomplete.

Earth is not a sealed spacecraft. It receives immense energy from the Sun, radiates energy back into space, exchanges some matter, and contains processes no crew designed. Its oceans and atmosphere are vastly larger than any engineered life-support loop. Its biological diversity, geologic cycles, and human societies cannot be governed from one control room.

But Earth is closed enough for consequences to return.

Carbon released in one place enters a shared atmosphere. A chemical designed for one use travels through water, air, food, organisms, and generations. Antibiotic pressure in one setting contributes to a resistance problem that crosses borders. Soil can lose function slowly and then fail quickly. Biodiversity can disappear before its control work is understood. A pathogen can move around the planet faster than the institution responsible for recognizing it.

The Earthrise lesson is not that the planet is a machine.

It is that there is no outside.

The control fantasy and the control obligation

A book with the word *control* in its title must end by confronting the oldest danger in the idea.

Human beings repeatedly mistake the ability to influence a system for the ability to command it. We dam rivers, suppress insects, redesign crops, administer drugs, alter behavior, build markets, and write algorithms. Many of these interventions produce extraordinary benefit. Some also move risk into a place or time where the original decision-maker no longer sees it.

The Universal Control Layer is not a claim that sufficient data will eliminate surprise. It is the opposite claim: surprise is permanent, so every consequential intervention needs feedback, boundaries, verification, and the humility to update the model.

The architecture is universal only in the questions it asks.

What outcome matters?

What state must be known?

What can be measured without destroying the evidence?

What does drift look like before failure?

Where is the highest-leverage point we can govern safely?

What is the least destructive effective action?

How will the system answer back?

Who has authority to act, and who bears the risk when the model is wrong?

Those questions belong in a neonatal screening program, a hospital water system, an agricultural field, a spacecraft, a pharmaceutical pipeline, and a planetary treaty. The answers should not be centralized into one model or one institution.

A human body is not a small Earth. A farm is not a patient. A city is not a space station. The same architecture can help each system without pretending that one set point or one controller should govern them all.

Distributed control is not a compromise forced by complexity. In living systems, it is often the only responsible design.

The ten lessons, reduced to one discipline

The stories in this book crossed a century of engineering and medicine.

Biosphere 2 taught that an inventory is not a state.

Fordlandia taught that removed context may have been performing invisible control.

Legionnaires' disease taught that the built environment participates in biology.

PKU screening taught that the useful sensor is defined by the intervention window.

The NASA Twins Study taught that baseline is a trajectory, not a photograph.

Mir and the International Space Station taught that a habitat and its crew form a coupled system.

The Panama Canal taught that the highest-leverage actuator may sit upstream of the obvious task.

HIV combination therapy taught that adaptive networks require more than one vulnerable point and that measurement is part of treatment.

Recombinant insulin taught that information becomes capability only through a qualified process.

Apollo 7 taught that the operator is inside the loop.

Taken together, they reduce to one discipline:

> ***Measure for purpose, modulate for purpose, verify for truth, and learn before the system forces the lesson upon you.***

The sequence matters.

Measurement without purpose becomes surveillance or accumulation.

Purpose without measurement becomes intention.

Modulation without understanding becomes force.

Verification without a willingness to change becomes ceremony.

Learning without memory becomes repetition.

The chemistry that makes the current move

The first book in this trilogy argued that biology's moving chemical layer remains radically undercounted. That argument matters here because control requires access to state.

A neuron does not fire because electricity exists as an independent substance. It fires because ion gradients have been created and maintained, channels change conformation, membranes separate charge, energy drives pumps, and molecular events reset the system for the next signal. A heart rhythm, an immune response, a metabolic transition, a wound, a memory, and a microbial negotiation all depend on changing molecular conditions.

Stop the molecular movement and the electrical description goes dark with it.

This does not make every molecule a controller. It means the chemical state is inseparable from the operation.

Much of modern measurement still favors what is known, stable, abundant, and easy to annotate. The off-peak feature, transient conjugate, spatial gradient, mixed-species product, or condition-specific molecular family often falls outside the decision. The instrument may see a trace that the library cannot name. The clinical report reduces the ocean to the few markers with accepted interpretation.

That is appropriate when an unvalidated feature could mislead care.

It is insufficient as a research strategy.

Industrial-Scale Microbiome Mining is one proposed way to expand the reference engine by recovering physical chemistry from intact communities at scales that make rare molecules experimentally available. It is not the only route and does not itself prove function. Its contribution is material and context: a larger set of real structures, source relationships, fractions, spectral patterns, and biological priors with which models and experiments can work.

The deployed future is different from the terrestrial mining platform. A patient or habitat needs a bounded monitor informed by a far larger reference system. The edge device does not need to carry the ocean. It needs to know which changes matter for the purpose at hand, how confidently they can be interpreted, and what action is allowed.

The reference engine learns broadly.

The deployed controller acts narrowly.

Confusing those roles would produce either a useless monitor or an unsafe one.

The era after generation becomes cheap

Artificial intelligence changes the economics of proposing answers.

Models can generate candidate molecules, explanations, targets, experiments, and control policies at a speed no human team can match. That capability is valuable. It also shifts scarcity.

When generation becomes cheap, grounded relevance becomes expensive.

A model can propose a million structures. It cannot make all of them biologically real. It can infer a network. It cannot assume the network survived sampling. It can predict a control action. It cannot erase the consequences of an unmeasured state. It can compress what is known and search patterns humans miss. It remains dependent on the quality, scope, provenance, and base rates of the data it receives.

The Universal Control Layer gives AI a disciplined job.

Not omniscient commander.

Interpreter under constraint.

It should identify anomalies, compare trajectories, preserve uncertainty, suggest confirmatory measurements, simulate bounded alternatives, expose assumptions, and help humans recognize when the present model no longer fits. Its authority should expand only where performance, safety, explainability, fallback, and governance justify expansion.

The most advanced model in the habitat should still be able to say: *This state is outside my validated experience. I recommend containment, additional measurement, and human review.*

Knowing when not to act is part of intelligence.

The smallest planet

Every long-duration spacecraft will be the smallest planet its crew has ever inhabited.

It will have atmosphere, water, carbon, nitrogen, surfaces, waste, food webs, microbes, machines, rules, memories, and people. Nothing thrown away will be truly away. A gram of material will matter because mass was paid for at launch. A failed crop will matter because the next harvest cannot be purchased down the road. A contaminated water loop will matter because the crew cannot open a new watershed. A conflict will matter because no one can leave the community.

The habitat will make visible what Earth allows us to ignore.

Waste is feedstock in the wrong place.

Resilience is capacity that looks idle until the day it is needed.

Diversity is not decoration when common failure modes exist.

Maintenance is biology when moisture, nutrients, and materials interact.

Health is environmental when air, food, light, work, and microbes shape the person.

Governance is life support when authority must move faster than communication.

A Mars habitat will not be saved by one miraculous microbial library, one perfect AI, one universal medicine, or one master control panel. It will survive through layered competence: robust mechanical systems, carefully managed biological processes, trained people, reference knowledge built on Earth, local sensing, conservative fallbacks, qualified interventions, and a culture capable of telling the truth when the state departs from the plan.

The same is true of Spaceship Earth, except the crew is larger, the purposes conflict, and no single mission director can define success.

What better control looks like on Earth

The off-world imperative earns attention because failure becomes immediate. The Earthbound work matters more because billions of people are already inside the loop.

A better control architecture in medicine would detect meaningful personal drift earlier, label uncertainty honestly, and connect intervention to verified function rather than a single convenient marker.

In public health, it would treat buildings, water, animals, climate, movement, and trust as components of the disease system rather than background around a pathogen.

In agriculture, it would monitor reserve capacity, soil function, ecological redundancy, and propagation risk rather than optimizing only annual output.

In pharmaceutical discovery, it would improve the priors feeding AI, recover natural chemistry that current libraries miss, design selective network interventions, and migrate from wide exploration to rigorous validation as exposure rises.

In environmental management, it would favor local, reversible, monitored actions over grand interventions whose effects cannot be contained or attributed.

In organizations, it would treat schedule, incentives, information flow, and fear as actuators capable of changing technical outcomes.

None of this requires waiting for a complete map.

It requires knowing which decisions the incomplete map can support.

A personal return to the instrument

I began the first book with the instruments my father installed and the lesson they gave me: the hidden becomes actionable only after an instrument makes it visible.

The trilogy ends one step later.

An instrument can reveal a shadow, a peak, a trajectory, a leak, a lesion, or a molecule. The next question determines whether the observation becomes benefit.

What will we do with what we can now see?

My career has repeatedly placed me at the boundary where a measurement system becomes a manufacturing process, where a process becomes a platform, where a platform is asked to become an asset, and where an asset is asked to justify the people and capital required to carry it forward. At every boundary, the temptation is to confuse the last achievement with the next capability.

Seeing is not controlling.

Controlling is not curing.

Generating is not validating.

Scaling is not governing.

The Universal Control Layer is an attempt to name the missing architecture between them.

I do not expect the final architecture to belong to one company or one discipline. It will be assembled from metabolomics, microbiology, medicine, ecology, control theory, materials science, synthetic biology, data engineering, AI, manufacturing, ethics, law, and the lived knowledge of operators and patients. Its most important advances may arrive from tools that do not yet share a vocabulary.

The work is not to force them into one institution.

The work is to make the interfaces capable of carrying truth.

And yet it answers back

Earthrise showed a planet without visible borders.

The control problem begins by restoring the borders that matter: the membrane, tissue, room, water loop, farm, watershed, habitat, jurisdiction, and decision right. It then asks how signals and consequences cross them.

The smallest planet we build will answer every design decision with chemistry.

The largest planet we inhabit is doing the same.

We do not need the fantasy of complete control.

We need a better relationship with feedback.

We need the ability to recognize state before failure, to intervene without destroying the regulator, to verify benefit rather than celebrate motion, and to change our own behavior when the living system gives us an answer we did not expect.

The final controller is not the algorithm.

It is the discipline to listen.

APPENDIX A

The Universal Control Layer Canvas

The canvas converts the book's architecture into one page of disciplined questions. It is not a substitute for a technical plan, clinical protocol, environmental assessment, or safety case. Its purpose is to expose what a team is assuming before sensors, models, and interventions become expensive.

Use it at the beginning of a program, at each major change in state or scope, and after any surprise that the current model did not predict.

1. Purpose

What human or system outcome are we trying to protect or improve?

Write the outcome in terms of function or benefit, not technology.

Weak: Develop an AI microbiome platform.

Stronger: Detect loss of potable-water safety early enough to prevent crew exposure and preserve the water loop.

Weak: Create a multi-target molecule.

Stronger: Restore a defined disease-relevant network state at clinically achievable exposure without unacceptable off-target toxicity.

Record:

- intended beneficiary;
- primary benefit;
- cost of delay;

- burden bearers;
- boundaries on acceptable action;
- what would count as completion.

2. State

Which variables together describe the condition that matters?

Do not begin with what the available instrument can measure. Begin with what the purpose requires knowing.

Consider:

- physical state;
- chemical state;
- biological composition;
- biological activity;
- spatial location;
- trajectory over time;
- reserve capacity;
- human experience and function;
- environmental and organizational context.

Distinguish a **state variable** from a **proxy**. Temperature may be a state variable in one control problem and only a proxy in another. A species count may be useful but insufficient when activity is the concern. A biomarker may move while patient function does not.

3. Sensing

How can the required state be measured at the right place, time, resolution, and confidence?

For each signal, specify:

- sampling location;
- frequency;

- delay from state to result;
- analytical confidence;
- detection limits;
- known interferences;
- effect of sampling on the system;
- personal or habitat baseline;
- confirmatory method;
- failure mode and calibration plan.

A sensor that reports too late for the intervention window is not an operational sensor, even if it is scientifically accurate.

4. Interpretation

What change is normal adaptation, recoverable drift, or danger?

Define:

- expected variation;
- direction and rate of change;
- uncertainty range;
- alert threshold;
- action threshold;
- confidence needed at each threshold;
- competing explanations;
- conditions outside the validated model.

Require the model to distinguish "unknown" from "safe."

5. Modulation

What is the least destructive effective action?

List interventions at several levels:

- observation only;
- additional measurement;

- behavioral or procedural change;
- environmental control;
- physical containment;
- ecological adjustment;
- nutritional intervention;
- drug or molecular intervention;
- combination intervention;
- manufacturing or supply action;
- emergency stabilization.

For each, state expected mechanism, time to effect, reversibility, collateral effects, control debt, and stop conditions.

6. Verification

How will the system demonstrate that the action helped?

Specify the chain:

action → mechanism → intermediate state → meaningful benefit.

Do not declare success at the earliest measurable point unless that point is the actual purpose.

Verification should include:

- intended response;
- unintended response;
- time window;
- negative control or counterfactual where possible;
- criteria for escalation, repetition, reversal, or stopping;
- human-reported function when relevant.

7. Learning

What changes after the result?

Record:

- updated baseline;
- updated model;
- revised thresholds;
- new unknowns;
- data and sample return;
- changes to the intervention library;
- changes to authority or fallback;
- evidence that should be shared beyond the program.

A closed loop that does not update its model is automation, not learning.

8. Safety and authority ring

The seven questions sit inside a ring of constraints.

- Who may observe?
- Who may recommend?
- Who may act?
- What actions are automated?
- What requires a second check?
- What requires consent?
- What is prohibited?
- What is the safe fallback?
- How can the person affected contest the interpretation?
- How will cybersecurity, data integrity, and model failure be handled?

Completed example: habitat water loop

Purpose. Maintain microbiologically safe potable water without exhausting stored reserves or damaging the recovery system.

State. Flow, stagnation time, temperature, disinfectant residual, conductivity, organic load, particles, pressure, biofilm indicators, targeted organisms, metabolic activity, crew symptoms, and material condition.

Sensing. Continuous physical and chemical sensors; scheduled targeted assays; triggered sampling after stagnation or maintenance; periodic reference chemistry; preserved samples for Earth analysis.

Interpretation. Separate sensor drift from true system drift. Use trajectory and spatial patterns. Flag low-flow zones before bulk-water counts rise. Treat unfamiliar chemical patterns as a request for confirmation, not proof of danger.

Modulation. Flush a local branch, change temperature or flow, isolate a segment, adjust validated disinfectant bounds, replace a component, switch to reserves, or initiate emergency treatment.

Verification. Confirm recovery at the point of use, not only at the treatment inlet. Verify both microbial and chemical safety and watch for rebound.

Learning. Update the map of vulnerable locations, material interactions, maintenance timing, and mission baseline.

APPENDIX B

State Vector and Baseline Worksheet

A single value rarely describes the state of a living system. A **state vector** is the smallest set of variables that together support a decision for a defined purpose.

The phrase should not be used to make a simple dashboard sound mathematical. The vector earns its name only when the variables are interpreted together and when their relationship to action is explicit.

Step 1: Write the decision

What decision must the state vector support?

Examples:

- Is this person's metabolic trajectory moving toward a clinically consequential state?
- Is a habitat water loop losing resilience?
- Is a microbial community producing chemistry associated with material degradation?
- Is a candidate network therapy engaging the intended targets at realistic exposure?

A vector designed for classification may be inadequate for control. "High risk" does not specify what action will change the trajectory.

Step 2: Build six layers

Layer 1: Outcome and function

What does the person or system need to be able to do?

Examples: maintain cognition, oxygenation, crop yield, barrier integrity, potable water, material strength, or viral suppression.

Layer 2: Primary controlled variables

Which variables are directly connected to the desired function?

Examples: glucose, oxygen partial pressure, viral load, corrosion rate, pressure, nutrient concentration.

Layer 3: Drivers and disturbances

What pushes the controlled variables?

Examples: meals, workload, infection, humidity, medication, flow, radiation, temperature, diet, microbial activity.

Layer 4: Reserve and resilience

How much capacity remains before the system loses control?

Examples: oxygen reserve, water inventory, renal clearance, immune reserve, battery capacity, ecological redundancy, spare parts, operator attention.

Layer 5: Context and boundary conditions

Where and when is the measurement meaningful?

Examples: tissue compartment, post-meal period, sleep phase, stagnant pipe branch, wound edge, crop-growth stage, mission phase.

Layer 6: Unknowns and confidence

Which variables are unmeasured, weakly identified, model-dependent, or outside the validated range?

A blank field is better than a fabricated certainty.

Step 3: Establish three baselines

Population baseline. How does the value compare with a relevant reference group?

Personal or system baseline. How does it compare with this individual, habitat, or process under a known stable condition?

Mission or purpose baseline. What range supports the particular function under the current task and environment?

These baselines can disagree. A value may be population-normal but abnormal for one astronaut. A value may differ from Earth baseline but represent stable adaptation in flight. A habitat may operate safely at a state that would be unacceptable in a hospital.

Step 4: Add time

For each variable, record:

- sampling interval;
- physiological or environmental lag;
- expected rate of change;
- intervention window;
- recovery time;
- seasonal, circadian, meal-related, or mission-related pattern.

The state vector is a movie. A dashboard that displays only the current frame should show direction and confidence beside the value.

Step 5: Add relationships

Ask:

- Which variables should move together?
- Which should move in opposite directions?
- Which change first?
- Which are compensatory responses rather than causes?
- Which relationship would falsify the current model?

The most informative signal may be a broken relationship rather than an extreme value.

Step 6: Define action classes

For each state pattern, choose one of five classes.

1. **Continue and observe.** Within expected range and trajectory.
2. **Increase sensing.** Uncertain or early drift; no intervention yet.
3. **Reversible adjustment.** Low-consequence action within validated bounds.
4. **Contain and escalate.** Potentially dangerous state or unfamiliar condition.
5. **Emergency action.** Delay is more dangerous than bounded intervention.

State-vector worksheet

Purpose:

Decision supported:

Outcome/function variables:

Controlled variables:

Drivers/disturbances:

Reserve/resilience variables:

Context/boundaries:

Unknowns:

Population baseline:

Personal/system baseline:

Purpose baseline:

Expected trajectory:

Alert pattern:

Action pattern:

Verification pattern:

Model-failure signal:

APPENDIX C

The Signal Evidence Ladder

The ladder prevents an unknown signal from being promoted directly into a diagnosis or intervention. It also prevents the opposite mistake: deleting every signal that lacks a complete name.

Rung 0: Unresolved observation

The instrument or observer records something outside expectation.

Examples: an unknown mass-spectrometry feature, unusual odor, unexplained fatigue, image anomaly, pressure fluctuation, unexpected microbial growth.

Permitted action: preserve the observation and context; check obvious artifacts.

Not yet justified: causal claim, diagnosis, product claim, high-consequence intervention.

Rung 1: Analytical reproducibility

The signal recurs under defined quality controls and is distinguishable from blanks, contamination, instrument drift, processing artifacts, and known analytical variants.

Questions:

- Does it repeat in the same sample?
- Does it recur in independent collections?
- Does an orthogonal method detect a related signal?
- Is the chain of custody intact?

Permitted action: monitor, fractionate, prioritize for identification, design follow-up.

Rung 2: Identity or bounded family

The signal is assigned to a confirmed structure, organism, material change, or well-bounded family with an explicit confidence level.

For chemistry, separate:

- feature;
- molecular formula;
- tentative structure;
- confirmed structure;
- stereochemical identity;
- authentic-standard match.

Permitted action: targeted measurement and mechanistic hypotheses.

Rung 3: Contextual association and plausible exposure

The identified signal changes with a state, location, time, exposure, disease, or intervention, and the amount can plausibly reach the relevant compartment.

Association is not causation. It is stronger when it is:

- replicated;
- dose- or trajectory-related;
- spatially coherent;
- temporally ordered;
- robust to major confounders;
- supported by a plausible mechanism.

Permitted action: stratification in research, confirmatory studies, low-risk monitoring.

Rung 4: Perturbation response

Changing the suspected source, pathway, signal, or environment changes the outcome in a relevant model.

Examples:

- add the molecule;
- remove the producer;
- inhibit the enzyme;
- change flow or humidity;
- restore the signal after removal;
- compare close analogs;
- test in an orthogonal model.

Permitted action: controlled experimental modulation; development of a causal model.

Rung 5: Causal and translational chain

Evidence supports the sequence from intervention through mechanism and intermediate state to meaningful outcome at realistic exposure.

Required elements vary by consequence but can include:

- necessity or strong causal contribution;
- sufficiency under defined conditions;
- relevant dose and compartment;
- specificity or understood network activity;
- reproducibility outside the originating group;
- safety and reversibility;
- manufacturability or operational feasibility.

Permitted action: staged clinical, environmental, or operational validation.

Rung 6: Qualified action

The signal and intervention have been validated for a defined use, population, environment, operating range, and decision authority.

Qualification includes:

- performance limits;
- false-positive and false-negative consequences;
- human factors;
- fallback;
- cybersecurity where relevant;
- labeling and training;
- monitoring after deployment.

Permitted action: use within the qualified envelope.

Rung 7: Learning deployment

Real-world use continuously checks whether the signal and action remain valid across populations, environments, time, and system changes.

The system looks for:

- model drift;
- distribution shift;
- rare harms;
- new confounders;
- resistance or adaptation;
- unequal access or performance;
- opportunities to narrow or retire the intervention.

A mature control layer can move a signal down the ladder when evidence weakens.

Confidence labels for chemical features

Use a simple label beside every important feature.

- **F:** reproducible feature only;
- **MF:** bounded molecular family;
- **T:** tentative structure;
- **C:** structure confirmed by authentic material or equivalent evidence;

- **X:** exposure in relevant compartment established;
- **A:** association replicated;
- **P:** perturbation response demonstrated;
- **Q:** qualified for a defined action.

The label keeps the unknown visible without pretending that it is known.

APPENDIX D

The Intervention Ladder and Control Target Brief

The strongest actuator is not automatically the best actuator. The ladder begins with observation and moves toward increasingly direct and consequential intervention.

Level 0: Preserve and observe

Use when the signal is uncertain, the system is stable, and additional observation is more valuable than action.

Risks: delay, missed intervention window.

Controls: explicit review date and escalation trigger.

Level 1: Improve measurement

Repeat, relocate, increase frequency, use an orthogonal method, or preserve context more effectively.

Risks: measurement burden, false reassurance from more data.

Controls: tie each new measurement to a decision.

Level 2: Behavioral or procedural adjustment

Change schedule, hygiene, diet, workflow, maintenance, or exposure.

Risks: low adherence, hidden burden, blame placed on the individual.

Controls: usability, consent, and verification.

Level 3: Environmental or physical control

Change flow, humidity, temperature, ventilation, light, spacing, isolation, filtration, or material condition.

Risks: energy cost, secondary ecological effects, shifting the problem.

Controls: monitor the downstream state and control debt.

Level 4: Ecological or nutritional modulation

Alter community resources, competitors, substrates, diet, prebiotic inputs, or biological process conditions.

Risks: context dependence, slow response, unintended community reorganization.

Controls: bounded trials, spatial monitoring, restoration plan.

Level 5: Targeted molecular intervention

Use a drug, biologic, peptide, small molecule, phage, inhibitor, agonist, or replacement factor against a defined mechanism.

Risks: off-target effect, exposure failure, metabolism, resistance, individual variability.

Controls: evidence ladder, dose, compartment, and response verification.

Level 6: Coordinated network intervention

Use a combination regimen, selective polypharmacological agent, multi-component ecological strategy, or synchronized environmental and molecular action.

Risks: interaction complexity, attribution difficulty, compounded toxicity, operational burden.

Controls: design brief, realistic exposure, component contribution, stop rules.

Level 7: Emergency stabilization

Use a broad, immediate action to prevent irreversible harm when the model is incomplete.

Examples: isolate a habitat loop, switch to reserves, provide empirical antimicrobial coverage, increase oxygen, abort a mission phase.

Risks: collateral damage and substantial control debt.

Controls: narrow time window, preauthorized conditions, rapid source investigation, restoration plan.

Stabilize, control source, restore

Every intervention plan should distinguish three purposes.

Stabilize. Prevent immediate failure.

Control the source. Interrupt the process producing the state.

Restore. Rebuild function and resilience.

One action may serve more than one purpose, but the team should not assume that stabilization removed the source or that source control restored the system.

Control Target Brief

Purpose and beneficiary:

Observed state and confidence:

Primary target:

Target type: environmental / behavioral / ecological / molecular / organizational / manufacturing.

Why this is the leverage point:

Expected mechanism:

Time to effect:

Realistic exposure or operating condition:

Reversibility:

Expected collateral effects:

Control debt created:

Verification chain:

Stop condition:

Fallback:

Authority and consent:

Learning return:

Selective-polypharmacology design questions

When the intervention intentionally engages several targets, add:

- Which targets are intended?
- Which are required, helpful, neutral, or unacceptable?
- What target engagement is achievable at one realistic exposure?
- Is the relationship simultaneous, sequential, spatial, or state-dependent?
- Does one interaction dominate toxicity?
- How does the pattern compare with close analogs?
- What network phenotype should change?
- How will promiscuity be distinguished from selectivity?
- What resistance or rerouting pathways remain?
- Can the design be simplified without losing the coordinated effect?

APPENDIX E

The ISMM Reference Engine and the Deployed Monitor

Industrial-Scale Microbiome Mining and a fielded monitoring system perform different jobs.

The reference engine learns broadly

The reference engine may include:

- industrial-scale recovery of intact environmental communities;
- source and environmental metadata;
- extraction and fractionation;
- high-resolution mass spectrometry;
- molecular networking and structural elucidation;
- genomics, transcriptomics, proteomics, and metabolomics;
- phenotype and target assays;
- authentic standards and synthesis;
- material-interaction studies;
- clinical and environmental cohorts;
- negative results and confidence labels;
- computational models and curated knowledge.

ISMM contributes where physical scale enables access to low-abundance chemistry and repeated experiments. It does not by itself establish human function, safety, or therapeutic value.

The deployed monitor acts narrowly

A hospital, farm, spacecraft, or patient device needs:

- a defined purpose;
- a bounded state vector;
- rugged assays;
- low mass, power, and consumables;
- rapid interpretation;
- known error modes;
- validated action thresholds;
- local fallback;
- human authority;
- secure updates.

The deployed monitor may use only a tiny fraction of the reference engine's knowledge.

The translation chain

1. **Discover.** A reproducible feature, family, organism, or state is observed.
2. **Identify.** Identity and context are established to the required confidence.
3. **Relate.** Association, exposure, and trajectory are characterized.
4. **Perturb.** Causal contribution is tested.
5. **Reduce.** The complex reference measurement is converted into a deployable signature or assay.
6. **Validate.** Performance is tested in the intended environment and population.
7. **Govern.** Authority, fallback, consent, and update rules are defined.
8. **Deploy.** The bounded monitor operates locally.
9. **Return.** Data and samples improve the reference engine.

Data-lineage requirements

Every deployable signal should retain links to:

- source material;
- collection conditions;
- analytical method;
- feature-processing history;
- identification confidence;
- reference standard;
- biological evidence;
- model version;
- deployment validation;
- known limitations;
- update history.

A model output without lineage cannot be responsibly promoted into control authority.

What belongs on Earth

- wide discovery;
- expensive structural work;
- large reference cohorts;
- broad model training;
- independent replication;
- high-complexity manufacturing development;
- long-term archival and governance.

What belongs at the edge

- purpose-specific sensing;
- personal or habitat baseline;
- anomaly recognition;

- confirmatory testing;
- bounded interpretation;
- reversible local action;
- safe fallback;
- compressed communication;
- evidence preservation.

The deployment test

Before a reference insight becomes a monitor, ask:

- Can the signal be measured reliably with edge constraints?
- Does it arrive before the intervention window closes?
- Does the action improve a meaningful outcome?
- Are false alarms and missed detections tolerable?
- Is the system robust to distribution shift?
- Can a user understand the recommendation?
- Can the system fail safely?
- Can the monitor be maintained without recreating the Earth laboratory?

APPENDIX F

Closed-Loop Risk Register and Autonomy Tiers

A closed-loop controller creates risks that an ordinary monitor does not. It can act repeatedly, amplify an error, and change the very state on which its model depends.

Risk categories

Purpose risk

The system optimizes the wrong outcome or a proxy that becomes detached from benefit.

Example: normalizes one biomarker while impairing function.

Sensor risk

Bias, drift, delay, location error, contamination, sampling disturbance, calibration failure, or adversarial input.

Model risk

Wrong causal structure, missing variable, distribution shift, overconfidence, unrecognized unfamiliar state.

Actuator risk

Wrong dose, wrong site, mechanical failure, delayed effect, saturation, interaction with another intervention.

Coupling risk

The action improves one subsystem while destabilizing another.

Human-factor risk

Alarm fatigue, poor usability, automation bias, gaming, concealment, training failure, conflict over authority.

Governance risk

Unclear accountability, unauthorized use, mission creep, inadequate consent, inaccessible appeal.

Cybersecurity risk

Corrupted sensor data, malicious update, stolen biological design, denial of service, altered manufacturing file.

Equity risk

The system works poorly for underrepresented populations or distributes benefit and burden unfairly.

Recovery risk

No safe fallback, insufficient reserves, failure to preserve evidence, inability to reverse action.

Risk-register fields

Hazard:

Initiating condition:

Affected purpose:

Likelihood and uncertainty:

Severity and irreversibility:

Detection method:

Preventive control:

Mitigating control:

Fallback:

Owner:

Trigger for review:

Residual risk:

Evidence required to close:

Autonomy tiers

Tier 0: Observe

The system collects, quality-checks, and displays data.

Appropriate when evidence is early, actuation is high consequence, or human expertise is immediately available.

Tier 1: Advise

The system recommends additional measurement, inspection, or consideration. It does not change the system.

Required: evidence display, confidence, alternatives, and user acknowledgment.

Tier 2: Protect reversibly

The system takes bounded, reversible actions such as increasing sampling, isolating a local loop, switching to a redundant sensor, or reducing workload.

Required: validated trigger, visible action, easy override, and automatic return criteria.

Tier 3: Modulate within a qualified envelope

The system adjusts a physiological, environmental, or manufacturing variable continuously within tested bounds.

Required: robust sensor validation, controller testing, cybersecurity, human factors, fallback, and post-action monitoring.

Tier 4: Emergency action

The system takes a high-consequence action because delay is expected to produce greater irreversible harm.

Required: narrowly predefined conditions, redundant evidence where possible, protected records, after-action review, and explicit authority.

Autonomy migration rules

Move upward only when:

- the purpose and controlled variable are stable;
- sensor performance is validated;
- the model's operating envelope is known;
- actuator performance and delay are characterized;
- fallback works;
- users understand the system;
- real-world monitoring supports the move.

Move downward when:

- sensors disagree;
- the state is unfamiliar;
- the model is updated materially;
- a critical component degrades;
- human trust or understanding falls;
- a rare harm appears;
- cybersecurity is uncertain.

The ability to reduce autonomy is a core safety function.

APPENDIX G

Local Biomanufacturing Charter

This charter is for a bounded product considered for manufacture outside a conventional centralized facility. It does not create regulatory authorization.

1. Product definition

- molecular identity;
- intended indication or function;
- dose and route;
- formulation;
- critical quality attributes;
- acceptable variants;
- prohibited substitutions.

2. Why local production

- mass saved;
- shelf-life advantage;
- uncertain demand;
- resupply delay;
- emergency value;
- terrestrial access benefit;
- strategic resilience.

Compare local production with stored inventory, not with an imaginary zero-cost alternative.

3. Production system

- living-cell, cell-free, chemical, enzymatic, or hybrid process;
- raw materials and consumables;
- power, water, gas, and waste;
- environmental requirements;
- operator steps;
- process sensors;
- cleaning and changeover;
- expected yield and cycle time.

4. Quality release

Define tests for:

- identity;
- purity;
- potency;
- sterility or microbial limits;
- endotoxin or process contaminants;
- concentration;
- formulation;
- stability;
- device or container compatibility.

No product should be released because the process completed without an alarm.

5. Update authority

Identify who may change:

- documentation;
- analytical software;
- process parameters inside a validated range;

- raw-material source;
- sequence or molecular design;
- dose or indication.

Higher biological consequence requires higher evidence and authority.

6. Failure and fallback

Plan for:

- contamination;
- low yield;
- wrong identity;
- failed quality test;
- sensor loss;
- cyber compromise;
- operator error;
- power interruption;
- waste-system failure;
- adverse patient response.

Fallback may be stored product, an alternative therapy, a second production path, remote consultation, or mission-level triage.

7. Patient and habitat loop

After use, monitor:

- intended response;
- pharmacokinetics where feasible;
- adverse effects;
- interaction with other products;
- environmental release;
- waste and decontamination;
- need for repeat dosing or reversal.

8. Learning return

Preserve:

- process data;
- quality results;
- retained sample;
- product version;
- patient state before and after;
- deviations;
- corrective actions;
- recommendations for the Earth reference process.

Release decision

Release approved by:

Evidence reviewed:

Known deviations:

Residual risk:

Monitoring plan:

Fallback ready: yes / no

APPENDIX H

Claims, Rights, and Ethics Covenant

The power to measure and modulate living systems must be governed before deployment, not after the first abuse.

Claim discipline

Every public or technical claim should be labeled as one of four types.

1. **Documented event.** A historical or observed fact.
2. **Established science.** A conclusion supported by a strong body of evidence.
3. **Engineering synthesis.** A proposed architecture or interpretation connecting established findings.
4. **Future scenario.** A plausible but unproved application.

Do not use commercial evidence as a substitute for independent scientific evidence. Do not describe a predicted molecule as a discovered medicine. Do not describe an association as a controller. Do not describe a prototype as a qualified system.

Purpose limitation

Data and intervention authority should be used only for the purpose under which they were collected or explicitly reauthorized.

Medical safety data should not silently become employment performance data.

Habitat surveillance should not become unrestricted personal surveillance.

Environmental monitoring should not become a license for unaccountable central control.

Consent and standing

People affected by the system should have meaningful standing proportional to the consequence.

That can include:

- informed consent;
- community consultation;
- patient representation;
- operator participation;
- protected dissent;
- appeal and correction;
- access to understandable explanations.

Emergency conditions can narrow consent but should not erase accountability.

Data rights

Define:

- ownership and stewardship;
- access;
- retention;
- secondary use;
- de-identification limits;
- cross-border transfer;
- security;
- right to correction;
- conditions for deletion;
- obligations after program closure.

Biological data can reveal information about family members and communities as well as the sampled person.

Minimum necessary intervention

Use the least destructive action capable of achieving the purpose within the intervention window.

This principle does not require timid action. In an emergency, the least destructive effective action may be aggressive. It requires comparing consequence, reversibility, control debt, and alternatives.

Dual-use review

Ask how the same capability could be used to:

- design toxins or pathogens;
- manipulate behavior;
- target populations;
- evade detection;
- alter ecosystems;
- create coercive surveillance;
- concentrate economic or political power.

Dual use does not automatically prohibit research. It changes access, disclosure, safeguards, and governance.

Parity by design

Measure performance across relevant populations and environments. Track who receives benefit, who bears false alarms, who lacks access, and whose data trained the system.

A control layer that widens avoidable disparity has failed part of its purpose even if its average accuracy is high.

Human review and contestability

High-consequence decisions require:

- evidence traceability;
- confidence and alternatives;
- qualified human review when available;
- the ability to stop or override;
- an auditable record;
- a path to challenge the result.

The covenant

Before deployment, the responsible organization should be able to state:

> *We will not claim more than the evidence supports. We will measure only for a defined purpose, act only within justified authority, preserve the right to challenge the model, use the least destructive effective intervention, verify benefit and harm, protect the data and the people who generate it, and change the system when reality contradicts our design.*

APPENDIX I

The Trilogy Crosswalk

The three books are designed as a progression, not three versions of the same argument.

Book 1: The Ghost Genome

Core question: What are we missing?

Primary domain: discovery science.

Owns:

- the hidden chemical output of living systems;
- cryptic assembly across species, host, diet, environment, and time;
- natural products and finite biological priors;
- the measurement challenge;
- Industrial-Scale Microbiome Mining as one route to physical access;
- ten stories about maps expanding after instruments changed.

Does not own: a general management system or a deployed control architecture.

Book 2: Breakthrough Aggressiveness

Core question: How should we organize work when the problem and proof burden differ?

Primary domain: innovation leadership and organizational design.

Owns:

- the seven-level Breakthrough Aggressiveness Scale;
- matching velocity and rigor to the problem;
- dynamic migration;

- BAS × TRL;
- BACS;
- professional comfort zones;
- investor and public accountability.

Does not own: the scientific case for the Ghost Genome or the specific architecture of biological sensing and modulation.

Book 3: The Universal Control Layer

Core question: What do we build when the map becomes useful and the organization can move?

Primary domain: systems integration and responsible operation.

Owns:

- purpose-bound state measurement;
- personal and habitat baselines;
- the signal evidence ladder;
- ISMM-informed monitoring;
- intervention and verification;
- selective network modulation;
- local biomanufacturing;
- autonomy, authority, fallback, ethics, and off-world use.

Does not re-teach: the ten discovery stories or the BAS scale.

The sequence in one line

Map the missing language. Engineer the way we advance it. Build the feedback systems that can use it responsibly.

Shared terms with different jobs

Ghost Genome. In Book 1, the object of discovery. In Book 3, a possible source of reference signals, interventions, and design priors.

ISMM. In Book 1, a discovery instrument. In Book 3, part of the Earth reference engine—not the deployed monitor.

AI. In Book 1, a model that needs better biological priors. In Book 2, an organizational mirror and decision aid. In Book 3, a bounded interpreter inside a control architecture.

Polypharmacology. In Book 1, a feature of natural chemical language and drug discovery. In Book 3, a design and control challenge requiring exposure, selectivity, verification, and safety.

Aggressiveness. In Book 2, the central operating variable. In Book 3, a mission posture that must respond to the biological and organizational state.

Recommended reading order

The books stand alone but accumulate.

Readers interested primarily in biology can begin with *The Ghost Genome*.

Leaders and investors can begin with *Breakthrough Aggressiveness*.

Readers focused on systems, diagnostics, space, or implementation can begin here, using the Preface and this crosswalk to understand the prior premises.

The complete trilogy asks one escalating question:

Can we discover more of what life is doing, organize ourselves to advance it correctly, and build systems wise enough to act without confusing influence with mastery?

Glossary

Actuator

The component or action that changes the system: a pump, valve, drug, schedule, environmental adjustment, ecological intervention, or human decision.

Adaptive system

A system whose behavior changes in response to experience, selection, learning, or changing conditions. The response can alter the effectiveness of future control actions.

Alert threshold

A value or pattern that triggers increased attention or measurement but not necessarily intervention.

Analytical confidence

The strength of evidence that a measurement corresponds to a real, reproducible signal and, where claimed, to a specific identity.

Autonomy tier

A defined level of authority granted to a system, from observation only through emergency action.

BAS

Breakthrough Aggressiveness Scale – The appropriate operating intensity level applied to a situation (problem).

Baseline

A reference state for comparison. It can be population-based, personal, system-specific, or purpose-specific and should include time and context.

Bayesian

A statistical framework that uses Bayes' Theorem to update the probability of a hypothesis as more evidence and data becomes available.

Biological software

A metaphor for the changing molecular, ionic, metabolic, and signaling state through which living machinery is coordinated. It is not literal computer code.

Closed loop

A system in which the result of an action is measured and used to adjust future action.

Control debt

A future burden created by an intervention that solves an immediate problem while reducing resilience, selecting resistance, shifting risk, or requiring later restoration.

Control point

A location in a causal process where an intervention can materially change the trajectory.

Cryptic assembly

The proposed or observed creation of molecules through transformations distributed across organisms, host processes, diet, environment, and time, such that no single genome directly specifies the final functional entity.

Deployed monitor

A purpose-specific sensing and interpretation capability operating in a hospital, habitat, field, or device. It uses a bounded subset of larger reference knowledge.

Digital twin

A computational representation intended to predict relevant behavior of a physical or biological system. A useful twin must preserve the variables and relationships needed for the decision.

Distribution shift

A change in the real-world population, environment, or state that makes model-training data less representative of current use.

Evidence ladder

The progression from unresolved observation through analytical reproducibility, identity, association, perturbation, causal chain, qualified action, and learning deployment.

Fallback

A safer or more conservative operating mode used when the primary system, sensor, model, or intervention is unreliable.

Feature

An analytical signal, often defined by mass-to-charge ratio, retention time, intensity, and related attributes. A feature is not automatically a unique molecule.

Ghost Genome

The book's term for the hidden or poorly characterized chemical output space generated by living systems and their interactions, especially chemistry not readable directly from one genome.

Human factors

The design and study of how people actually use, understand, maintain, override, and respond to a system under real conditions.

Industrial-Scale Microbiome Mining (ISMM)

A proposed and developing approach to recover intact environmental microbial communities at industrial scale so low-abundance physical chemistry can become available for fractionation, analysis, and testing without requiring standard culture as the first gate.

Intervention window

The period during which action can still prevent or meaningfully alter the outcome.

Managed ecology

The deliberate maintenance of microbial or ecological function within safety bounds rather than the unrealistic goal of permanent total sterility.

Metabolic fingerprint

A pattern of measured metabolites or chemical features associated with a state. It can be useful before every component is identified but requires explicit confidence and validation.

Model envelope

The range of states, inputs, populations, and environments over which a model has been validated.

Modulation

A controlled change intended to alter a trajectory. It can be physical, environmental, behavioral, ecological, molecular, or organizational.

Molecular family

A group of related chemical features or structures supported by fragmentation, biosynthetic logic, mass differences, retention behavior, or other evidence.

Off-peak or unannotated signal

A measured feature not included in the familiar set of interpreted markers. It may be noise, artifact, known chemistry in an unfamiliar form, or a genuine unknown; it requires the evidence ladder.

Open loop

An action taken without measuring and using the response to adjust subsequent action.

Personal baseline

An individual's own reference trajectory under defined conditions, used alongside population ranges.

Physiologic closed-loop control

Automated adjustment of a therapy or physiologic variable using feedback from a sensor, control logic, and an actuator.

Polypharmacology

The interaction of one drug or regimen with multiple biological targets. It can be intentional and selective or unintended and promiscuous.

Purpose baseline

The state range required for a defined task or outcome, which may differ from a population or Earth reference range.

Qualified process

A manufacturing or operational process demonstrated to produce a defined output reliably within specified conditions and controls.

Reference engine

The large Earth-based knowledge and material system that integrates discovery collections, analytical data, structural confirmation, biological evidence, clinical and environmental cohorts, and model updates.

Reserve capacity

The remaining ability of a system to absorb disturbance and recover before function is lost.

Resilience

The capacity to maintain or restore meaningful function after disturbance. Stability can exist temporarily while resilience is being consumed.

Selective polypharmacology

Intentional, coordinated engagement of a bounded set of targets or pathways at realistic exposure to produce a desired network effect while limiting unwanted interactions.

Sensor

A device, assay, observation, or human report that provides information about the system state.

Set point

A desired target value in a controller. Complex living systems often require ranges, trajectories, or multiple objectives rather than one fixed set point.

Signal evidence label

A concise marker of what is known about a signal, from reproducible feature through qualified action.

Stabilization

Immediate action to keep a system within survivable or functional bounds. Stabilization may not remove the source or restore resilience.

State

The condition of a system at a particular time and context, represented by variables sufficient for the defined purpose.

State vector

The smallest practical set of variables interpreted together to support a defined decision.

Systems biology

The study of biological components as interacting networks and dynamic systems rather than isolated parts alone.

Universal Control Layer

The proposed architecture described in this book: measure a living system for a defined purpose, interpret drift in context, modulate at an appropriate point, verify the result, and update the model within explicit safety and governance boundaries.

Verification

Measurement of whether an intervention produced the intended meaningful result without unacceptable harm elsewhere.

World model

The representation a controller uses to explain and predict the system. It can include causal relationships, baselines, constraints, and uncertainty.

NOTES AND SOURCES

A note on method

The narratives in this book are reconstructions from public records, primary scientific papers, official reports, oral histories, and responsible secondary accounts. They are written as stories, but the story does not license invented dialogue or certainty. Where historians disagree, where a later retelling has simplified a complicated event, or where a scientific result is suggestive rather than causal, the manuscript preserves that distinction.

The notes below are not intended to reproduce every source consulted. They identify the records that carry the principal factual load, explain important editorial choices, and point readers toward deeper treatment. Claims about the proposed Universal Control Layer remain the author's engineering synthesis unless otherwise stated.

Author's Note and Preface

The control-loop language draws on a long intellectual history rather than one modern technology. Foundational works include Walter B. Cannon's account of physiological regulation in *The Wisdom of the Body* (1932); Norbert Wiener's *Cybernetics: Or Control and Communication in the Animal and the Machine* (1948); W. Ross Ashby's *An Introduction to Cybernetics* (1956); and later systems-biology syntheses such as Hiroaki Kitano, "Systems Biology: A Brief Overview," *Science* 295 (2002): 1662-1664, doi:10.1126/science.1069492. The book's seven-question architecture is original to this manuscript, though it plainly stands inside that lineage.

The statement that electrical signaling depends on molecular conditions is standard physiology, not a claim that neurons are merely chemical. Action potentials depend on ion gradients, membrane channels, pumps, cellular energy, and neurotransmitter release and clearance. The point is integrative: electrical and chemical descriptions are complementary views of the same living event.

For modern physiologic closed-loop technology, see U.S. Food and Drug Administration, *Technical Considerations for Medical Devices with Physiologic Closed-Loop Control Technology: Guidance for Industry and Food and Drug Administration Staff* (2023). The guidance is narrower than the architecture proposed here, but it is a useful example of the safety, human-factors, sensor, algorithm, actuator, and fallback questions that arise when a controller can act on a patient.

The Ghost Genome is defined consistently with the first book: not a second literal DNA sequence, but the poorly characterized chemical output space generated through host, microbial, dietary, environmental, and transient processes. The relevant background manuscript is *The Ghost Genome: The Hidden Chemical Language Inside Us - and Why Medicine Has Barely Begun to Read It*, third editor's story draft, July 2026.

Breakthrough Aggressiveness is used here only as a prior management premise: match operating intensity to the problem and migrate posture as evidence, maturity, reversibility, and time pressure change. The detailed BAS, BAS x TRL matrix, and BACS tools belong to *Breakthrough Aggressiveness: When to Be Dogged, When to Leap, and How to Make Innovation Deliver*, first-edit draft, July 2026.

Chapter 1 - The Oxygen That Disappeared

The first Biosphere 2 closure began September 26, 1991, and ended September 26, 1993. The eight crew members were Abigail Alling, Linda Leigh, Taber MacCallum, Mark Nelson, Jane Poynter, Sally Silverstone, Mark Van Thillo, and Roy Walford. The project combined scientific ambition, theatrical visibility, private financing, institutional conflict, and genuine whole-system research. Accounts that reduce it either to a flawless success or a fraudulent stunt miss what makes the event useful.

The principal oxygen-loss source is Jeffrey P. Severinghaus, Wallace S. Broecker, William F. Dempster, Taber MacCallum, and Martin Wahlen, "Oxygen Loss in Biosphere 2," *Eos, Transactions American Geophysical Union* 75, no. 3 (1994): 33, 35-37, doi:10.1029/94EO00285. The investigators reported that oxygen declined from near ambient levels to roughly 14 percent during the first sixteen months. Their evidence implicated microbial respiration of organic-rich soils. Carbon dioxide released through respiration did not accumulate in the amount a simple balance predicted because carbonation of exposed concrete removed substantial carbon dioxide from the atmosphere. The manuscript therefore distinguishes the oxygen-consuming process from the process that obscured part of its expected carbon-dioxide signature.

For the structure, atmospheric dynamics, soil-based life-support system, and lessons of the closures, see Mark Nelson, William Dempster, Norberto Alvarez-Romo, and Taber MacCallum, "Atmospheric Dynamics and Bioregenerative Technologies in a Soil-Based Ecological Life Support System: Initial Results from Biosphere 2," *Advances in Space Research* 18, no. 4-5 (1996): 151-176; and Mark Nelson, "Biosphere 2's Lessons about Living on Earth and in Space," *Space: Science & Technology* (2021): 8067539, doi:10.34133/2021/8067539.

Jane Poynter's memoir, *The Human Experiment: Two Years and Twenty Minutes Inside Biosphere Two* (Thunder's Mouth Press, 2006), and Mark Nelson's *Pushing Our Limits: Insights from Biosphere 2* (University of Arizona Press, 2018) provide participant perspectives. They should be read alongside technical literature because memory, faction, and later interpretation shaped the public story.

The phrase "largest sealed ecological experiment" is used in the practical sense of a materially closed, human-inhabited ecological facility at unprecedented scale. It is not a claim that every material exchange was literally zero. Biosphere 2 was unusually tight, not metaphysically isolated.

Chapter 2 - The Plantation That Invited Its Enemy

Ford Motor Company's Brazilian rubber projects operated from the late 1920s through 1945. Fordlandia was established along the Tapajos River; Belterra followed as a second plantation. The project attempted to secure a vertically integrated rubber supply while importing industrial, architectural, dietary, labor, and social assumptions from the United States.

The central narrative source is Greg Grandin, *Fordlandia: The Rise and Fall of Henry Ford's Forgotten Jungle City* (Metropolitan Books, 2009). Grandin treats the plantation as both an industrial project and an effort to impose a Fordian social order. John Galey, "Industrialist in the Wilderness: Henry Ford's Amazon Venture," *Journal of Interamerican Studies and World Affairs* 21, no. 2 (1979): 261-289, provides an earlier scholarly account.

The Henry Ford museum's research collection, "Ford Rubber Plantations in Brazil," documents the company's operations and dates. Museum of the Americas, "Fordlandia: The Lost City of Henry Ford," provides a concise visual and historical overview. These institutional accounts support the broad facts but do not replace Grandin's treatment of labor, ecology, and power.

The chapter's ecological argument requires nuance. Rubber trees did not possess a conscious "spacing strategy." In the native Amazonian forest, *Hevea brasiliensis* is dispersed among other species. Plantation concentration changed transmission opportunities for South American leaf blight and other pests. The manuscript uses "invisible spacing" as an engineering description of an ecological effect, not as teleology.

The worker revolt commonly called the Breaking Pans revolt occurred in December 1930 after accumulated grievances over food, labor discipline, timekeeping, cultural imposition, and management. It is used here as a feedback signal from a social system, not as proof that every management decision was biologically caused.

Fordlandia's failure had multiple causes: poor site and soil decisions, disease and pest pressure, lack of tropical agronomic expertise, logistics, labor conflict, changing economics, and the rise of synthetic rubber. The chapter resists the cleaner but false story that one fungus alone defeated Henry Ford.

Chapter 3 - The Hotel That Became a Habitat

The American Legion convention in Philadelphia ran in July 1976. The subsequent outbreak became one of the most publicized disease investigations in U.S. history. The Centers for Disease Control and Prevention's historical account reports 221 people sick and 34 deaths across the broader recognized outbreak.

The foundational epidemiologic paper is David W. Fraser et al., "Legionnaires' Disease: Description of an Epidemic of Pneumonia," *New England Journal of Medicine* 297 (1977): 1189-1197, doi:10.1056/NEJM197712012972201. Joseph E. McDade et al., "Legionnaires' Disease: Isolation of a Bacterium and Demonstration of Its Role in Other Respiratory Disease," *New England Journal of Medicine* 297 (1977): 1197-1203, doi:10.1056/NEJM197712012972202, describes isolation of the organism and its relationship to the outbreak and earlier respiratory disease.

For retrospective context, see William C. Winn Jr., "Legionnaires Disease: Historical Perspective," *Clinical Microbiology Reviews* 1, no. 1 (1988): 60-81. The CDC's David J. Sencer Museum, "Legionnaires' Disease," preserves the institutional history and public-pressure context.

The outbreak was associated with the Bellevue-Stratford Hotel, but the exact physical point of release was not conclusively demonstrated during the original investigation. Later knowledge established that *Legionella* can amplify in warm building-water systems and spread through aerosol-generating devices. The manuscript therefore treats the hotel as a habitat and amplification environment without asserting that the original investigators proved one specific cooling tower as the source.

The organism's relationship with free-living amoebae is now central to understanding its environmental persistence and intracellular biology. A useful review is Paul S. Hoffman, "Legionella in the Environment: The Hidden Life of an Accidental Pathogen," in the broad modern literature on *Legionella* ecology. The chapter's phrase "organism inside another organism" refers to this ecology, not to the discovery story's first hypothesis.

The discussion of unannotated mass-spectrometry features is supported by the metabolomics literature. See Wout B. Dunn et al. on reporting and identification standards; and Pieter C. Dorrestein and colleagues on molecular networking. A concise statement of the annotation problem appears in da Silva, Dorrestein, and

Quinn, "Illuminating the Dark Matter in Metabolomics," *Proceedings of the National Academy of Sciences* 112 (2015): 12549-12550, doi:10.1073/pnas.1516878112.

Chapter 4 - The Baby Who Looked Healthy

Phenylketonuria can be clinically silent in the newborn period while elevated phenylalanine damages the developing nervous system. Early dietary treatment can prevent much of the severe intellectual disability associated with untreated classical PKU. The chapter uses PKU because it makes the intervention window visible: the test matters because action before symptoms changes the life course.

Robert Guthrie and Ada Susi, "A Simple Phenylalanine Method for Detecting Phenylketonuria in Large Populations of Newborn Infants," *Pediatrics* 32 (1963): 338-343, PMID 14063511, is the foundational paper for the bacterial inhibition assay. The dried-blood-spot collection system made population-scale screening logistically practical.

For the development, implementation controversies, and persistence required to create newborn screening, see Harvey L. Levy, "Robert Guthrie and the Trials and Tribulations of Newborn Screening," *International Journal of Neonatal Screening* 7, no. 1 (2021): 5, doi:10.3390/ijns7010005; and K. Bhattacharya et al., "The Evolution of Blood-Spot Newborn Screening," *Translational Pediatrics* 3, no. 2 (2014): 63-70.

Guthrie's family history influenced his determination to make testing practical and widely available. Biographical accounts describe a son with developmental disability and a niece with PKU. The manuscript presents this as motivation reported in historical accounts, not as a private reconstruction of conversations.

Modern newborn screening uses multiple technologies, including tandem mass spectrometry, and varies by jurisdiction. Screening is not diagnosis. Positive screens require confirmation, and programs must manage false positives, false negatives, follow-up, treatment access, consent, data retention, and equity.

The chapter's extension from PKU to chemical early warning is an engineering analogy, not evidence that untargeted metabolomics is ready to diagnose unspecified disease from "noise." Unannotated features may be artifacts, adducts, fragments, contaminants, benign variation, consequences rather than causes, or useful signals. The Signal Evidence Ladder in Appendix C exists to prevent the metaphor from outrunning evidence.

Chapter 5 - One Twin Left Earth

Scott Kelly launched to the International Space Station in March 2015 and returned in March 2016 after roughly 340 days in space. His identical twin, Mark Kelly, served as a genetically matched ground comparator. The design was scientifically valuable and intrinsically limited: one flight subject and one comparison individual cannot establish population-wide causality.

The primary integrated report is Francine E. Garrett-Bakelman et al., "The NASA Twins Study: A Multidimensional Analysis of a Year-Long Human Spaceflight," *Science* 364, no. 6436 (2019): eaau8650, doi:10.1126/science.aau8650. The study integrated genomic, epigenomic, transcriptomic, proteomic, metabolomic, microbiome, immune, cardiovascular, cognitive, and other measurements.

The paper reported a mixture of transient and persistent changes. Many measures returned toward preflight values after landing, while some did not immediately. The chapter deliberately avoids the popular shorthand that spaceflight "rewrote" the astronaut or that every difference was harmful.

For the gut-microbiome findings, see Aleksandar A. Voorhies et al., "Study of the Impact of Long-Duration Space Missions at the International Space Station on the Astronaut Microbiome," *Scientific Reports* 9 (2019): 9911, doi:10.1038/s41598-019-46303-8. The changes in the Twins Study were measurable but generally modest and largely diminished after return. Diet, shared environments, sampling, and individual variability complicate interpretation.

The "personal baseline" discussion is consistent with longitudinal medicine and personalized monitoring but should not be confused with a fully validated digital twin. A digital twin in healthcare can range from a statistical personal model to a high-fidelity mechanistic simulation. The manuscript uses **adaptive counterpart** to emphasize uncertainty, recalibration, and bounded decision support rather than a perfect virtual duplicate.

The distinction between resilience and accommodation is an engineering synthesis. A system may maintain output while consuming reserve. Measuring reserve capacity remains domain specific; there is no universal resilience biomarker.

Chapter 6 - The Biology Behind the Panels

Microbial contamination has been documented throughout human spaceflight. In a closed habitat, crew-associated microorganisms meet unusual surfaces, humidity pockets, cleaning regimes, radiation, microgravity-related effects, limited resupply, and long maintenance cycles.

For Russian spacecraft and Mir, see Natalia D. Novikova, "Review of the Knowledge of Microbial Contamination of the Russian Manned Spacecraft," *Microbial Ecology* 47 (2004): 127-132, doi:10.1007/s00248-003-1055-2. Reports from Mir described microbial growth and material interactions in condensate-prone locations. Popular retellings sometimes turn this into "mutant space fungus ate the station." The evidence supports microbial colonization, material degradation concerns, and corrosion-associated processes; it does not support a simple single-cause account of Mir's many failures.

For culture-independent analysis of International Space Station surfaces, see Kasthuri Venkateswaran and colleagues, including Christine Checinska Sielaff et al., "Characterization of the Total and Viable Bacterial and Fungal Communities Associated with the International Space Station Surfaces," *Microbiome* 7 (2019): 50, doi:10.1186/s40168-019-0666-x.

Pratyay Sengupta et al., "Genomic, Functional, and Metabolic Enhancements in Multidrug-Resistant *Enterobacter bugandensis* Facilitating Its Persistence and Succession in the International Space Station Microbiome," *Microbiome* 12 (2024): 62, doi:10.1186/s40168-024-01777-1, analyzed ISS isolates and their genomic and functional distinctions. The study raises important questions about persistence, adaptation, and interactions in a closed habitat. It did not report that these strains caused astronaut disease. The manuscript states that distinction explicitly.

Planetary protection is governed through international and national policy, including the Committee on Space Research's *COSPAR Policy on Planetary Protection*. Forward contamination concerns material carried from Earth to another world; backward contamination concerns potentially harmful extraterrestrial material returned to Earth's biosphere. The policy evolves as missions and scientific understanding change.

The chapter's "sampling paradox" is a recurring microbiology problem: the act of removing an organism from its spatial, chemical, and ecological context can change the state being measured. Culture, sequencing, microscopy, spatial methods, metabolomics, and environmental sensors each see different projections. No one method preserves the complete habitat.

Chapter 7 - The Canal Medicine Built

The French canal effort in Panama began in the 1880s and failed amid engineering, financial, managerial, and disease catastrophes. The United States took control of the project in 1904 and completed the canal in 1914. Disease control was not the only reason for success, but without it the labor force and enterprise could not be sustained.

Credit for the mosquito-transmission insight must not begin with the United States. Cuban physician Carlos J. Finlay proposed in 1881 that a mosquito transmitted yellow fever. The U.S. Army Yellow Fever Commission led by Walter Reed later produced influential experimental evidence in Cuba. William C. Gorgas applied mosquito control and sanitation at scale in Havana and Panama.

Useful sources include John R. Pierce and James V. Writer, *Yellow Jack: How Yellow Fever Ravaged America and Walter Reed Discovered Its Deadly Secrets* (Wiley, 2005); David McCullough, *The Path Between the Seas: The Creation of the Panama Canal, 1870-1914* (Simon & Schuster, 1977); and Paul S. Sutter, "'The First Mountain to Be Removed': Yellow Fever Control and the Construction of the Panama Canal," *Environmental History* 21 (2016): 250-279.

The chapter's sequence - insight, confirmation, operational application, infrastructure - is intentionally broader than a hero story. Finlay, Reed's commission, laboratory personnel, nurses, sanitation workers, inspectors, engineers, local communities, and thousands of laborers all carried parts of the result.

The medical accomplishments existed inside colonial rule, racial hierarchy, unequal housing and treatment, coercive labor arrangements, and ethically unacceptable human experimentation by modern standards. See Johanna Braunold, "How Racism and 'Tropical Medicine' Built the Panama Canal," *AMA Journal of Ethics* 26, no. 2 (2024): E123-E128. The chapter includes this history because a control system that treats some people as expendable has not succeeded morally, even when the physical project is completed.

The "intervention ladder" and "highest-leverage point you can govern safely" are original tools in this manuscript. They are compatible with public-health prevention hierarchies and control theory but should not be read as clinical guidance.

Chapter 8 - The Cocktail That Bent the Curve

Combination antiretroviral therapy transformed HIV infection for people who could obtain and sustain treatment. The turning point was not one trial or one molecule; it emerged from virology, drug development, viral-load measurement, clinical trials, regulatory change, activist pressure, community knowledge, manufacturing, and access.

Roy M. Gulick et al., "Treatment with Indinavir, Zidovudine, and Lamivudine in Adults with Human Immunodeficiency Virus Infection," *New England Journal of Medicine* 337 (1997): 734-739, doi:10.1056/NEJM199709113371102, demonstrated durable viral suppression with a three-drug regimen relative to two drugs in the studied population.

Scott M. Hammer et al., "A Controlled Trial of Two Nucleoside Analogues plus Indinavir in Persons with Human Immunodeficiency Virus Infection and CD4 Cell Counts of 200 per Cubic Millimeter or Less," *New England Journal of Medicine* 337 (1997): 725-733, doi:10.1056/NEJM199709113371101, provided related clinical evidence.

Frank J. Palella Jr. et al., "Declining Morbidity and Mortality among Patients with Advanced Human Immunodeficiency Virus Infection," *New England Journal of Medicine* 338 (1998): 853-860, doi:10.1056/NEJM199803263381301, documented major declines in opportunistic disease and mortality associated with more intensive antiretroviral therapy in a large outpatient cohort.

Viral-load measurement changed the feedback loop by making response and rebound visible before many clinical endpoints. For the relationship between virologic markers and outcomes, see David A. Katzenstein et al., "The Relation of Virologic and Immunologic Markers to Clinical Outcomes after Nucleoside Therapy in HIV-Infected Adults," *New England Journal of Medicine* 335 (1996): 1091-1098.

Patient activism changed research priorities, trial access, endpoint debates, regulatory practice, communication, and the authority structure of the response. See Steven Epstein, *Impure Science: AIDS, Activism, and the Politics of Knowledge* (University of California Press, 1996); and Peter Staley, *Never Silent: ACT UP and My Life in Activism* (Chicago Review Press, 2021).

Combination therapy is not synonymous with selective polypharmacology. A regimen combines distinct agents. A polypharmacological molecule engages multiple targets. Both can produce coordinated network effects, but their pharmacology, controllability, resistance barriers, manufacturing, dosing, and safety differ.

The chapter challenges the automatic use of "promiscuous" as a dismissal, not the existence of promiscuous and toxic compounds. Natural origin does not confer purpose, safety, or selectivity. Concentration, free exposure, tissue distribution, metabolites, binding kinetics, off-targets, and context must be measured.

Chapter 9 - The Message That Became Medicine

Insulin was first used successfully in a person with diabetes in 1922. For decades, commercial insulin was purified from animal pancreases. That supply chain saved lives while creating constraints related to source material, purification, immunogenicity, consistency, manufacturing, and scale.

The recombinant-human-insulin story includes competing teams, synthetic-gene methods, biotechnology startups, established pharmaceutical manufacturing, and regulatory review. David V. Goeddel et al., "Expression in *Escherichia coli* of Chemically Synthesized Genes for Human Insulin," *Proceedings of the National Academy of Sciences* 76 (1979): 106-110, doi:10.1073/pnas.76.1.106, is a primary technical paper.

The U.S. Food and Drug Administration approved Humulin on October 28, 1982. FDA historical materials describe it as the first biosynthetic human insulin product and the first approved medical product derived from recombinant-DNA technology. The production system converted a defined information sequence into a biological product through a qualified manufacturing chain; it did not make manufacturing "just data."

For the broader development history, see Arthur D. Riggs, "Making, Cloning, and the Expression of Human Insulin Genes in Bacteria: The Path to Humulin," *Endocrine Reviews* 42, no. 3 (2021): 374-380, doi:10.1210/endrev/bnaa029.

The portable-manufacturing discussion draws on work such as Keith Pardee et al., "Portable, On-Demand Biomolecular Manufacturing," *Cell* 167 (2016): 248-259.e12, doi:10.1016/j.cell.2016.09.013, which demonstrated freeze-dried, cell-free systems capable of producing biomolecules after rehydration. NASA's BioNutrients program has explored in-space nutrient production using engineered biological systems. These are enabling precedents, not evidence that a Mars habitat can safely synthesize an arbitrary newly designed medicine on demand.

Local production requires product-specific identity, potency, purity, sterility or bioburden control, stability, release testing, equipment qualification, raw-material traceability, cybersecurity, update authority, and clinical decision rules. A data file cannot replace matter, energy, reagents, process knowledge, or quality systems.

The book's future scenario deliberately proceeds in stages: known stable products first; more complex biologics later; adaptive or newly designed therapeutics only after evidence and governance justify them. The Local Biomanufacturing Charter in Appendix G is a conceptual checklist, not a manufacturing protocol.

Chapter 10 - The Cold That Rewrote the Flight Plan

Apollo 7 launched October 11, 1968, and splashed down October 22. The crew was Walter M. Schirra Jr., Donn F. Eisele, and R. Walter Cunningham. It was the first crewed Apollo mission and the first human test of the redesigned command and service module after the Apollo 1 fire.

NASA mission reports, transcripts, press materials, and biomedical histories document the crew's respiratory illness, medication discussions, fatigue, schedule disputes, communications conflict, and successful completion of major test objectives. See Charles A. Berry, "Apollo 7 to 11: Medical Concerns and Results," NASA technical record (1969); Richard S. Johnston et al., eds., *Biomedical Results of Apollo* (NASA SP-368, 1975); and NASA, *Apollo 7 Mission Report*.

Schirra developed cold symptoms early in the mission; Eisele and Cunningham later reported symptoms. In microgravity, congestion and fluid shifts complicated comfort and the ability to clear pressure. The crew resisted wearing helmets during reentry partly because of concern that blocked ears could not be equalized. NASA ultimately accepted the crew's decision.

Popular history often calls Apollo 7 a "mutiny." That label creates narrative force but can flatten the context. The crew completed the mission's essential technical objectives. The conflict reflected illness, workload, personality, communications, command authority, post-Apollo-1 pressure, and differing assessments of what the plan required. The chapter uses the episode to examine how a schedule becomes an actuator inside a human system, not to diagnose the crew from a distance.

The Apollo 1 fire on January 27, 1967, killed Virgil "Gus" Grissom, Edward White II, and Roger Chaffee during a ground test. The redesigned spacecraft and NASA's institutional recovery shaped the pressure surrounding Apollo 7. For official background, see NASA's Apollo history and the Apollo 204 Review Board materials.

The chapter's autonomy tiers are an original synthesis informed by human factors, aviation, medicine, autonomous systems, and delayed-communication mission design. They require domain-specific development before use.

Epilogue - The Smallest Planet We Have Ever Built

Apollo 8 entered lunar orbit in December 1968. On December 24, William Anders photographed the Earth rising above the lunar horizon while Frank Borman and James Lovell were also aboard. The color image became known as *Earthrise*. Anders later emphasized the reversal at the heart of the mission: the crew went to explore the Moon and discovered the Earth.

For official mission history and imagery, see NASA's Apollo 8 mission archive and the NASA image catalog entry for AS08-14-2383. Robert Poole, *Earthrise: How Man First Saw the Earth* (Yale University Press, 2008), examines the image's cultural meaning.

Buckminster Fuller's *Operating Manual for Spaceship Earth* (1969) helped popularize the closed-planet metaphor. The manuscript uses "Spaceship Earth" as a boundary and responsibility metaphor, not as a claim that the planet can be governed like a spacecraft or optimized to one shared set point.

The "generation becomes cheap" argument applies to AI-generated hypotheses and structures: abundance of proposals can make relevance, provenance, causal evidence, and physical validation more scarce, not less. See Jack W. Scannell et al., "Diagnosing the Decline in Pharmaceutical R&D Efficiency," *Nature Reviews Drug Discovery* 11 (2012): 191-200, doi:10.1038/nrd3681; and Anthony D. King et al. on the limitations and opportunities of AI-enabled discovery. The exact economics will change; the base-rate problem remains.

Industrial-Scale Microbiome Mining is described as a proposed route to expand physical chemical priors by recovering intact environmental communities at scales that can make low-abundance chemistry available for fractionation and testing. Claims about its output, scale, and commercial promise require independent diligence and should remain separated from established peer-reviewed evidence.

Every story requires both a vision and the apparatus to build it. In bringing this book to life, I leaned fearlessly into modern tools—leveraging Large Language Models and the vast index of the internet as research partners, sounding boards, and synthesizers. While the soul, worldbuilding, and core concepts of this series are entirely my own, these technologies allowed me to test ideas, refine complex details, and build a more efficient, resonant narrative. To embrace the future of storytelling is not to replace human imagination, but to supercharge it.

One final note- These stories involved mentors of differing interactions, but lessons were learned because of those efforts, their work, and patience with me. Thank you, Guy Carter, Ph.D., Professor Pieter Dorrestein, Ph.D., & Jack Scannell, Ph.D., and David Newman, Ph.D. So many other could be thanked.

SELECTED BIBLIOGRAPHY

Control, systems, and complexity

Ashby, W. Ross. *An Introduction to Cybernetics*. Chapman & Hall, 1956.

Cannon, Walter B. *The Wisdom of the Body*. W. W. Norton, 1932.

Checkland, Peter. *Systems Thinking, Systems Practice*. Wiley, 1981.

FDA. *Technical Considerations for Medical Devices with Physiologic Closed-Loop Control Technology*. Guidance for Industry and FDA Staff, 2023.

Kitano, Hiroaki. "Systems Biology: A Brief Overview." *Science* 295 (2002): 1662-1664. doi:10.1126/science.1069492.

Meadows, Donella H. *Thinking in Systems: A Primer*. Chelsea Green, 2008.

Sterman, John D. *Business Dynamics: Systems Thinking and Modeling for a Complex World*. Irwin/McGraw-Hill, 2000.

Wiener, Norbert. *Cybernetics: Or Control and Communication in the Animal and the Machine*. MIT Press, 1948.

Biological chemistry, metabolomics, and networks

Dührkop, Kai, Huibin Shen, Marvin Meusel, Juho Rousu, and Sebastian Böcker. "Searching Molecular Structure Databases with Tandem Mass Spectra Using CSI:FingerID." *Proceedings of the National Academy of Sciences* 112 (2015): 12580-12585. doi:10.1073/pnas.1509788112.

Hopkins, Andrew L. "Network Pharmacology: The Next Paradigm in Drug Discovery." *Nature Chemical Biology* 4 (2008): 682-690. doi:10.1038/nchembio.118.

Newman, David J., and Gordon M. Cragg. "Natural Products as Sources of New Drugs over the Nearly Four Decades from 1981 to 2019." *Journal of Natural Products* 83 (2020): 770-803. doi:10.1021/acs.jnatprod.9b01285.

Quinn, Robert A., Louis-Felix Nothias, Oliver Vining, Michael Meehan, Edward Esquenazi, and Pieter C. Dorrestein. "Molecular Networking as a Drug Discovery, Drug Metabolism, and Precision Medicine Strategy." *Trends in Pharmacological Sciences* 38 (2017): 143-154.

da Silva, Ricardo R., Pieter C. Dorrestein, and Robert A. Quinn. "Illuminating the Dark Matter in Metabolomics." *Proceedings of the National Academy of Sciences* 112 (2015): 12549-12550. doi:10.1073/pnas.1516878112.

Wishart, David S. "Emerging Applications of Metabolomics in Drug Discovery and Precision Medicine." *Nature Reviews Drug Discovery* 15 (2016): 473-484.

Closed ecosystems and human spaceflight

Garrett-Bakelman, Francine E., et al. "The NASA Twins Study: A Multidimensional Analysis of a Year-Long Human Spaceflight." *Science* 364 (2019): eaau8650. doi:10.1126/science.aau8650.

Johnston, Richard S., Lawrence F. Dietlein, and Charles A. Berry, eds. *Biomedical Results of Apollo*. NASA SP-368, 1975.

Nelson, Mark. *Pushing Our Limits: Insights from Biosphere 2*. University of Arizona Press, 2018.

Nelson, Mark. "Biosphere 2's Lessons about Living on Earth and in Space." *Space: Science & Technology* (2021): 8067539. doi:10.34133/2021/8067539.

Novikova, Natalia D. "Review of the Knowledge of Microbial Contamination of the Russian Manned Spacecraft." *Microbial Ecology* 47 (2004): 127-132.

Poynter, Jane. *The Human Experiment: Two Years and Twenty Minutes Inside Biosphere Two*. Thunder's Mouth Press, 2006.

Severinghaus, Jeffrey P., Wallace S. Broecker, William F. Dempster, Taber MacCallum, and Martin Wahlen. "Oxygen Loss in Biosphere 2." *Eos* 75 (1994): 33, 35-37. doi:10.1029/94EO00285.

Sengupta, Pratyay, et al. "Genomic, Functional, and Metabolic Enhancements in Multidrug-Resistant *Enterobacter bugandensis* Facilitating Its Persistence and Succession in the International Space Station Microbiome." *Microbiome* 12 (2024): 62. doi:10.1186/s40168-024-01777-1.

Voorhies, Aleksandar A., et al. "Study of the Impact of Long-Duration Space Missions at the International Space Station on the Astronaut Microbiome." *Scientific Reports* 9 (2019): 9911. doi:10.1038/s41598-019-46303-8.

Medicine, surveillance, and intervention

Epstein, Steven. *Impure Science: AIDS, Activism, and the Politics of Knowledge*. University of California Press, 1996.

Fraser, David W., et al. "Legionnaires' Disease: Description of an Epidemic of Pneumonia." *New England Journal of Medicine* 297 (1977): 1189-1197. doi:10.1056/NEJM197712012972201.

Gulick, Roy M., et al. "Treatment with Indinavir, Zidovudine, and Lamivudine in Adults with Human Immunodeficiency Virus Infection." *New England Journal of Medicine* 337 (1997): 734-739. doi:10.1056/NEJM199709113371102.

Guthrie, Robert, and Ada Susi. "A Simple Phenylalanine Method for Detecting Phenylketonuria in Large Populations of Newborn Infants." *Pediatrics* 32 (1963): 338-343.

Hammer, Scott M., et al. "A Controlled Trial of Two Nucleoside Analogues plus Indinavir in Persons with Human Immunodeficiency Virus Infection and CD4 Cell Counts of 200 per Cubic Millimeter or Less." *New England Journal of Medicine* 337 (1997): 725-733.

Levy, Harvey L. "Robert Guthrie and the Trials and Tribulations of Newborn Screening." *International Journal of Neonatal Screening* 7 (2021): 5.

McDade, Joseph E., et al. "Legionnaires' Disease: Isolation of a Bacterium and Demonstration of Its Role in Other Respiratory Disease." *New England Journal of Medicine* 297 (1977): 1197-1203. doi:10.1056/NEJM197712012972202.

Palella, Frank J. Jr., et al. "Declining Morbidity and Mortality among Patients with Advanced Human Immunodeficiency Virus Infection." *New England Journal of Medicine* 338 (1998): 853-860. doi:10.1056/NEJM199803263381301.

Staley, Peter. *Never Silent: ACT UP and My Life in Activism*. Chicago Review Press, 2021.

Engineering, ecology, and infrastructure

Braunold, Johanna. "How Racism and 'Tropical Medicine' Built the Panama Canal." *AMA Journal of Ethics* 26 (2024): E123-E128.

Galey, John. "Industrialist in the Wilderness: Henry Ford's Amazon Venture." *Journal of Interamerican Studies and World Affairs* 21 (1979): 261-289.

Grandin, Greg. *Fordlandia: The Rise and Fall of Henry Ford's Forgotten Jungle City*. Metropolitan Books, 2009.

McCullough, David. *The Path Between the Seas: The Creation of the Panama Canal, 1870-1914*. Simon & Schuster, 1977.

Pierce, John R., and James V. Writer. *Yellow Jack: How Yellow Fever Ravaged America and Walter Reed Discovered Its Deadly Secrets*. Wiley, 2005.

Sutter, Paul S. "'The First Mountain to Be Removed': Yellow Fever Control and the Construction of the Panama Canal." *Environmental History* 21 (2016): 250-279.

Biotechnology, manufacturing, and the economics of discovery

Goeddel, David V., et al. "Expression in *Escherichia coli* of Chemically Synthesized Genes for Human Insulin." *Proceedings of the National Academy of Sciences* 76 (1979): 106-110. doi:10.1073/pnas.76.1.106.

Pardee, Keith, et al. "Portable, On-Demand Biomolecular Manufacturing." *Cell* 167 (2016): 248-259.e12. doi:10.1016/j.cell.2016.09.013.

Riggs, Arthur D. "Making, Cloning, and the Expression of Human Insulin Genes in Bacteria: The Path to Humulin." *Endocrine Reviews* 42 (2021): 374-380.

Scannell, Jack W., Alex Blanckley, Helen Boldon, and Brian Warrington. "Diagnosing the Decline in Pharmaceutical R&D Efficiency." *Nature Reviews Drug Discovery* 11 (2012): 191-200. doi:10.1038/nrd3681.

ACKNOWLEDGMENTS AND FIRST-RELEASE NOTE

This trilogy began with a recurring frustration: extraordinary knowledge can coexist with a missing way of seeing.

The Ghost Genome asked what biology has failed to count. *Breakthrough Aggressiveness* asked why capable organizations so often apply the wrong operating intensity to important work. This third book asks what becomes possible - and what new responsibilities appear - when better measurement and better innovation are connected through feedback.

The ideas in these pages owe their existence to many communities that rarely occupy the same room: microbiologists, natural-products chemists, metabolomics researchers, physicians, ecologists, control engineers, environmental scientists, manufacturing teams, data scientists, astronauts, mission planners, patients, advocates, public-health workers, and the people who maintain buildings, water systems, laboratories, farms, and factories after the visionary meeting is over.

I am especially grateful to the researchers whose work made the ten stories available for responsible retelling. The scientific papers and historical accounts listed in the notes do more than support facts. They preserve the intellectual chain by which an observation became a measurement, a measurement became an intervention, and an intervention became part of public capability.

Biosortia's work shaped the questions in this book. It taught me to think about scale as an instrument, physical chemistry as a data source, and the difference between a computational possibility and material that a living community actually made. It also taught me the obligation to distinguish ambition from validation. A commercial effort may illuminate a path without proving the destination.

The people who have worked with me across environmental engineering, manufacturing, entrepreneurship, and biotechnology contributed lessons that no citation can fully capture: a system fails at the interface; the operator usually sees drift before the dashboard; an accepted constraint may be an artifact of an old process; speed without purpose creates waste faster; and a measurement is not valuable merely because it is precise.

Several future scenarios are intentionally ambitious. They are included because a capstone book should show where the architecture could lead. They should remain visibly labeled as future scenarios until the required sensing, data, causal evidence, manufacturing, regulation, governance, and safety exist.

The deepest premise is more modest.

Every living system answers back.

Our responsibility is to measure what matters, act only with justified authority, and learn quickly enough to hear the answer before the system loses the ability to recover.